EXAM QUESTION PRACTICE PACK

AQA A-LEVEL GEOGRAPHY

HODDER EDUCATION
AN HACHETTE UK COMPANY

Every effort has been made to trace all copyright holders, but if any have been inadvertently overlooked, the Publishers will be pleased to make the necessary arrangements at the first opportunity.

Although every effort has been made to ensure that website addresses are correct at time of going to press, Hodder Education cannot be held responsible for the content of any website mentioned in this book. It is sometimes possible to find a relocated web page by typing in the address of the home page for a website in the URL window of your browser.

Hachette UK's policy is to use papers that are natural, renewable and recyclable products and made from wood grown in sustainable forests. The logging and manufacturing processes are expected to conform to the environmental regulations of the country of origin.

Orders: please contact Bookpoint Ltd, 130 Park Drive, Milton Park, Abingdon, Oxon OX14 4SE. Telephone: +44 (0)1235 827827. Fax: +44 (0)1235 400401. Email: education@bookpoint.co.uk. Lines are open from 9 a.m. to 5 p.m., Monday to Saturday, with a 24-hour message answering service. You can also order through our website: www.hoddereducation.co.uk.

ISBN: 978-1-5104-7711-7

First published in 2020 by Hodder Education,
An Hachette UK Company
Carmelite House
50 Victoria Embankment
London EC4Y 0DZ

www.hoddereducation.co.uk

Impression number 10 9 8 7 6 5 4 3 2

Year 2024 2023 2022 2021 2020

Cover photo: mariana_designer - stock.adobe.com

Typeset by Aptara, India

Printed by Hobbs the Printers Ltd, Totton, Hampshire SO40 3WX

A catalogue record for this title is available from the British Library.

FSC
www.fsc.org
MIX
Paper from responsible sources
FSC™ C104740

CONTENTS

INTRODUCTION

This pack of exam-style questions, mark schemes and example responses is specially curated for the AQA A-level Geography specification. The topics covered are:

- Physical geography
  - Section A Water and carbon cycles
  - Section B option: Coastal systems and landscapes
  - Section C option: Hazards
- Human geography
  - Section A Global systems and global governance
  - Section B Changing places
  - Section C option: Contemporary urban environments
  - Section C option: Population and the environment

The pack is divided into two sections:

- **Exam-style questions.** Two or three sets of questions for each section of each paper are presented in a similar way to the actual exam papers. You may wish to photocopy all or part of them for use with your class.
- **Mark schemes and example responses.** Each question has two student responses — a 'Student A' response typical of an answer that would receive high marks, and a 'Student B' response, which would receive fewer marks. Each response includes a commentary that describes why they receive the marks they do. The mark scheme for each question indicates how the responses could be graded, and can be used alongside each type of student response or just with the question.

This pack is designed to help you to:

- encourage students to reflect on their responses and ensure they know how to succeed
- cultivate students' key skills and knowledge by regular assessment throughout the course or in the revision period before the exams
- incorporate question practice into your lesson plans in the final, vital stage of teaching a topic: putting theory into practice
- teach flexibly, picking and choosing photocopiable pages as appropriate to share with students
- facilitate peer discussion of what is good or better about given answers, which allows greater insight into quality responses
- allow students to analyse responses without the bias that can come from looking at their own or their friends' work — and so get more from the task

Assessment objectives (AOs)

Assessment objectives are national criteria that are used by examiners when writing exam papers to ensure there is an even coverage of skills. They determine what students are being asked to do and what command terms and key words are being used in the question.

Assessment objective descriptors

Assessment objective	Description
AO1	Demonstrate knowledge and understanding of places, environments, concepts, processes, interactions and change, at a variety of scales
AO2	Apply knowledge and understanding in different contexts to interpret, analyse and evaluate geographical information and issues
AO3	AO3: Use a variety of relevant quantitative, qualitative and fieldwork skills to: • investigate geographical questions and issues • interpret, analyse and evaluate data and evidence • construct arguments and draw conclusions

Breakdown of assessment objectives for A-level

	AO1	AO2	AO3	Overall weighting
Component 1	15–20%	14–19%	1–6%	40%
Component 2	15–20%	14–19%	1–6%	40%
Component 3 (fieldwork investigation)	0%	2%	18%	20%
Total	**30–40%**	**30–40%**	**20–30%**	**100%**

EXAM-STYLE QUESTIONS

Paper 1 Physical geography

Resources

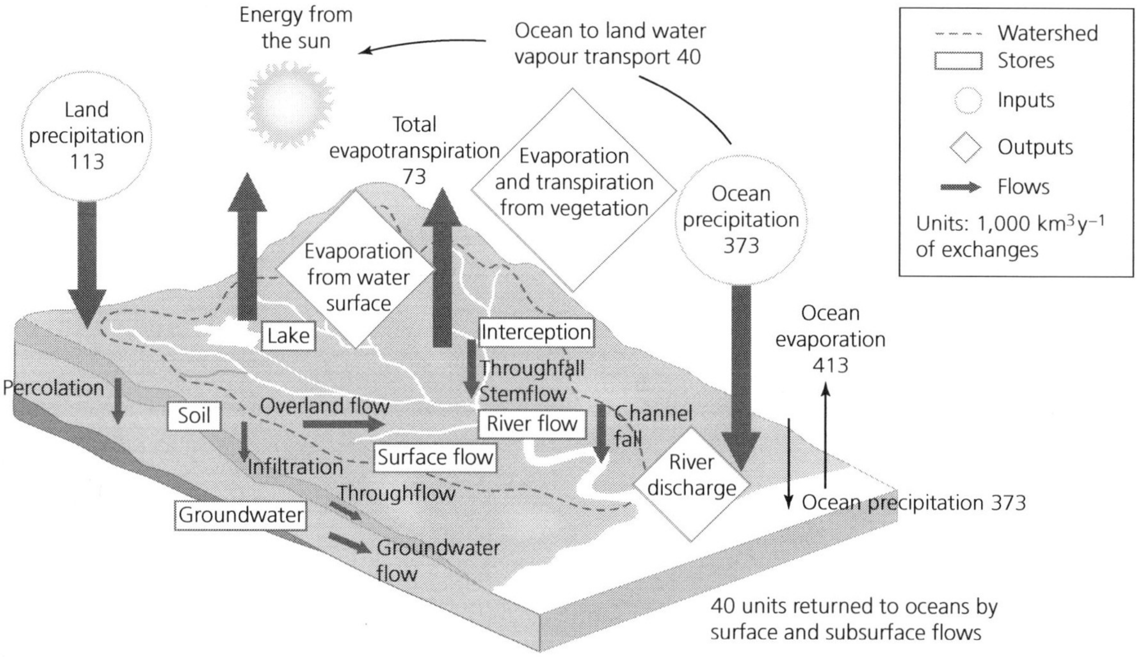

Figure 1 The drainage basin hydrological cycle

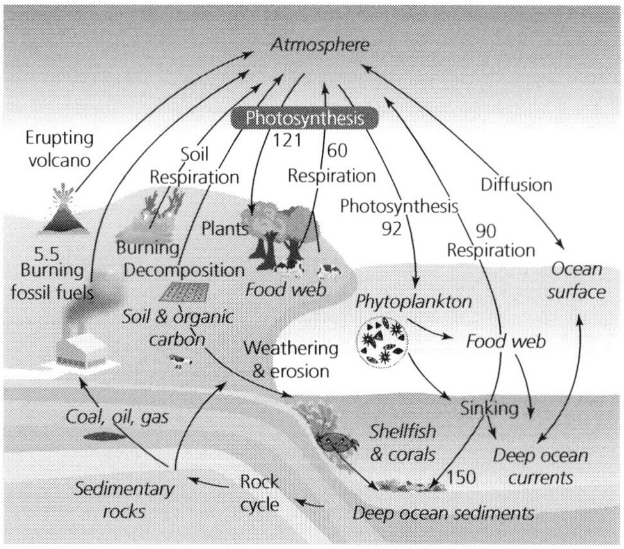

Figure 2 The carbon cycle. Reservoirs are shown in italic, other labels represent processes. Numbers represent transfer/flux in gigatonnes of carbon per year

Figure 3 The dry River Glen in Lincolnshire

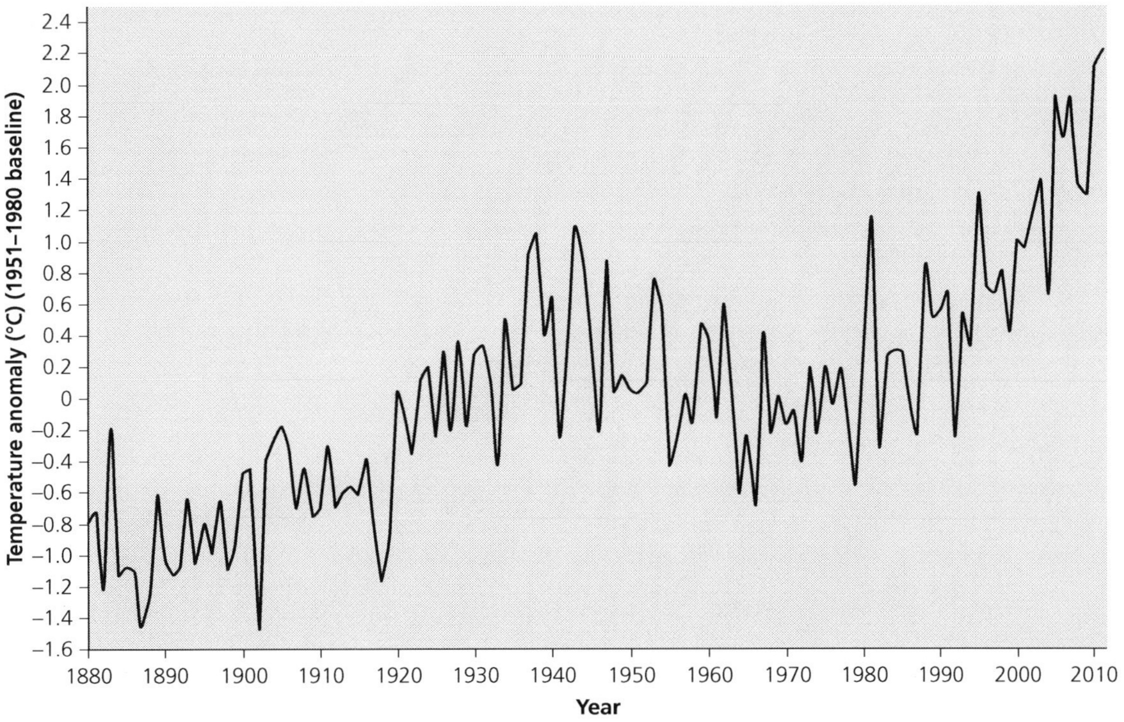

Figure 4 Arctic temperature change (°C) (64–90°N) (Source: data from NASA GISS)

If farmers continue to deforest the Amazon Basin, São Paulo could face devastating water shortages. The Amazon Basin acts as a 'biotic pump' that provides the energy for the 'rivers in the sky', which then flow towards São Paulo. Without the rainforest, the level of rainfall in southeast Brazil will drop dramatically. According to records stretching back to 1931, in 2014–15 the amount of rainfall was 50% lower than the previous worst year.

The effects of this lack of water supply reached the smart eateries of central São Paulo, where customers were forced to use plastic cutlery and eat off plastic plates as there was no water to run the dishwashers. There was not even enough water for Starbucks to make coffee, so it could serve only bottled beer and canned soft drinks.

Although the water company has planted 45,000 hectares of trees, benefiting in the process from climate regulation, it has no influence on Amazonian conservation. In addition, illegally built housing around a major reservoir has resulted in water too contaminated to use.

Incentives to reduce consumption of water were offered to customers, and people have continued to use less despite the current availability of water. Large industrial clients were also offered discounts to discourage them from sinking their own wells. There is a need for more investment in the forests and in water treatment, rather than in dams and reservoirs, to build climate resilience for the future.

Figure 5 Article on water shortages in São Paulo

Figure 6 Cliffs in East Sussex

Degrees of freedom	Significance level	
	0.05	**0.01**
6	0.82	0.94
7	0.71	0.89
8	0.64	0.83
9	0.60	0.78

Figure 7 Critical values for R_s

Figure 8 The Green Bridge of Wales

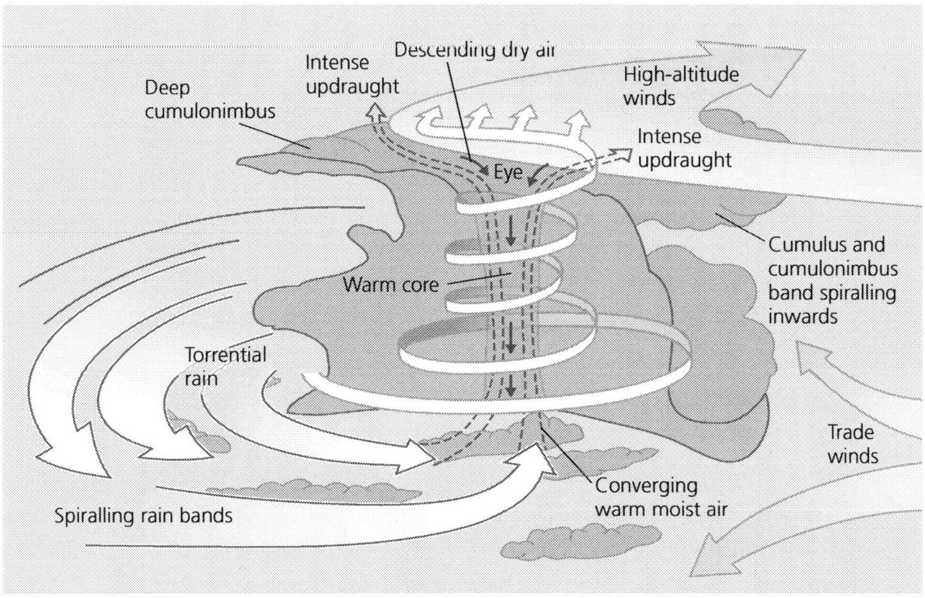

Figure 9 The structure of a tropical revolving storm

Figure 10 Flooding in Temerloh, Malaysia, December 2013

Figure 11 Flooding in New Port Richey, Florida, June 2012

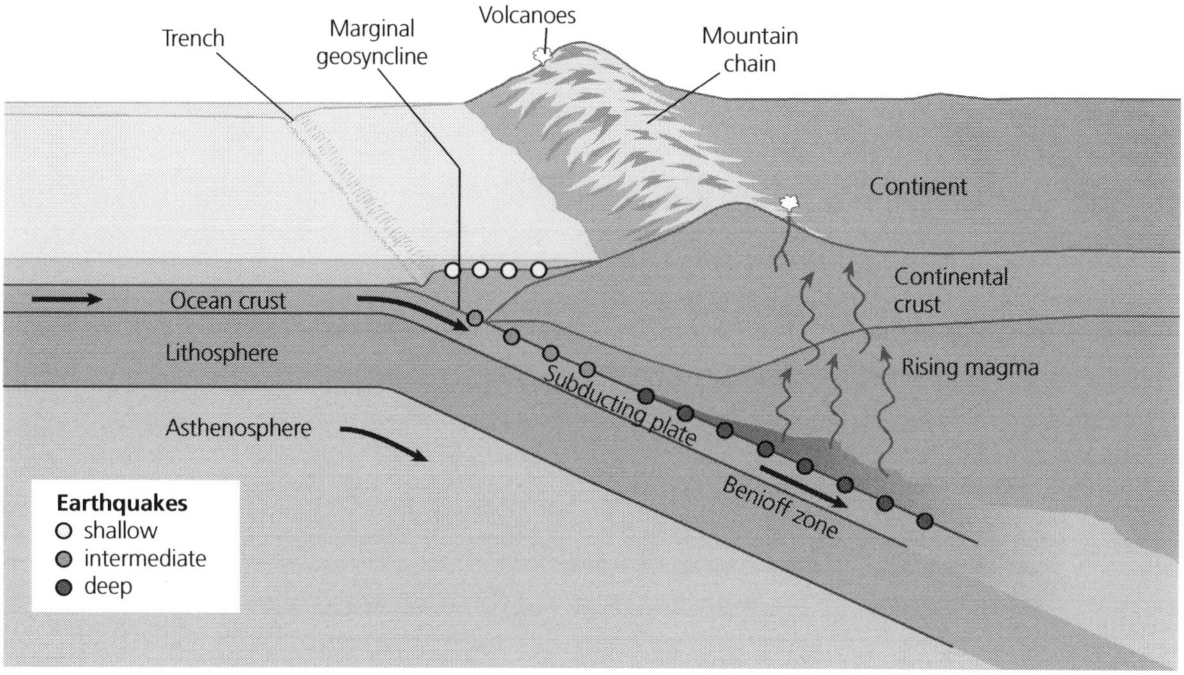

Figure 12 Cross-section of a plate boundary

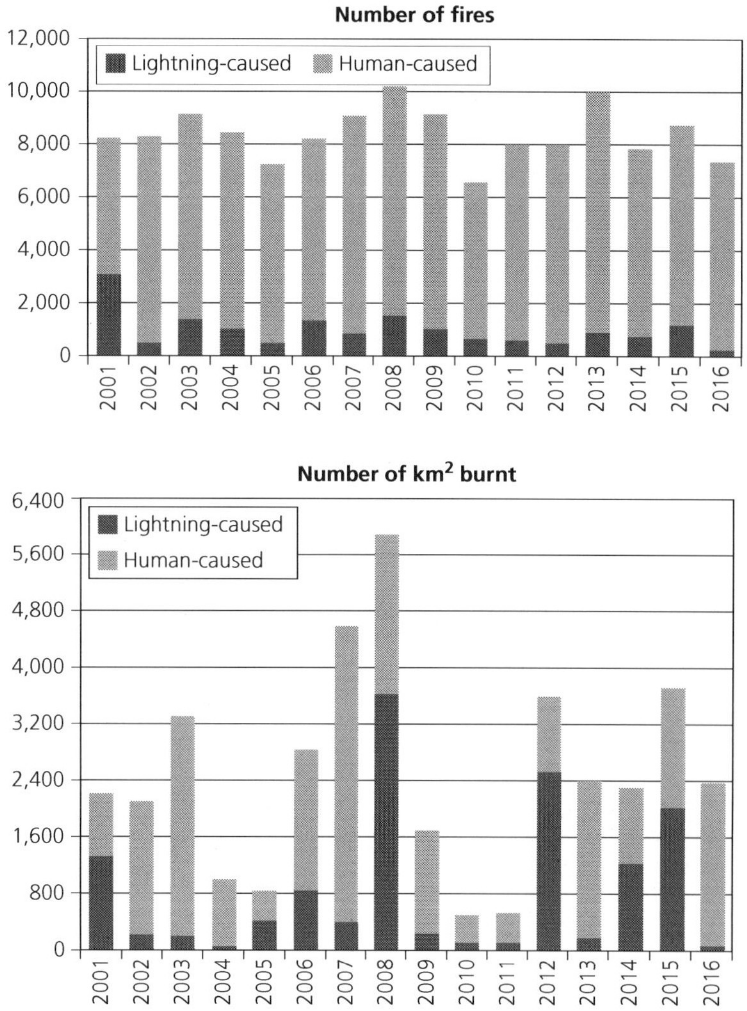

Figure 13 Lightning- and human-caused fires in California (Source: data from National Interagency Coordination Center).

Section A Water and carbon cycles

Set 1

Question 1 Flow

Explain the concept of 'flow' as it applies to a drainage basin. **(4 marks)**

(Mark scheme and example answers on page 79)

Question 2 Groundwater

Using **Figure 1** (see page 6) and your own knowledge, outline how climate change might affect the levels of groundwater in the UK. **(6 marks)**

(Mark scheme and example answers on page 80)

Question 3 The carbon cycle

Using **Figure 2** (see page 6), explain how changes in the carbon cycle over time may have an impact on global climate. **(6 marks)**

(Mark scheme and example answers on page 82)

Question 4 The importance of vegetation

Assess the importance of vegetation in linking the water and carbon cycles. **(20 marks)**

Total marks: 36

(Mark scheme and example answers on page 83)

Set 2

Question 5 Feedback

Explain **either** 'positive feedback' **or** 'negative feedback'. **(4 marks)**

....

....

....

....

....

....

....

....

(Mark scheme and example answers on page 87)

Question 6 Carbon dioxide and climate change

Table 1 shows data produced by the US National Oceanic and Atmospheric Administration. Complete the table and calculate the Spearman's rank correlation coefficient (R_s) using the formula provided.

Interpret your answer using the information provided. **(6 marks)**

Null hypothesis: There is no relationship between global temperature change and average atmospheric carbon dioxide.

Table 1 A comparison between global temperatures and average atmospheric carbon dioxide (Source: NOAA)

Year	Average atmospheric temperature compared with long-term average temperature (1900–2010) (°C)	Rank R_1	Average CO_2 measured at Mauna Loa Observatory, Hawaii (ppm)	Rank R_2	Difference in ranks (d)	d^2
1970	0	9.5	328	10	−0.5	0.25
1975	0	9.5	330	9	0.5	0.25
1980	0.3	7	338	8	−1	1
1985	0.1	8	346	7	1	1
1990	0.4	5.5	353	6	−0.5	0.25
1995	0.5		361			
2000	0.4	5.5	369	4	1.5	2.25
2005	0.7	2.5	379	3	−0.5	0.25
2010	0.7	2.5	390	2	0.5	0.25
2015	0.9	1	401	1	0	0

$\Sigma d^2 =$ ______________________

Formula for calculating Spearman's rank correlation coefficient:

$$R_s = 1 - \left(\frac{6 \sum d^2}{n^3 - n} \right)$$

$R_s =$ ______________________

where R_S is Spearman's rank correlation coefficient, *n* is the number of variables and *d* is the difference in the ranks of the variables.

For 10 pairs of variables, the critical values for R_S are ±0.564 at the 5% significance level and ±0.746 at the 1% significance level.

....

....

....

....

(Mark scheme and example answers on page 88)

Question 7 Soil water budget

Using **Figure 3** (see page 6) and your own knowledge, assess the possible changes to the soil water budget in southern England in a warming climate. **(6 marks)**

....

....

....

....

....

....

....

....

....

....

....

....

(Mark scheme and example answers on page 89)

Question 8 Demand for energy and the carbon cycle

'The demand for energy in the modern world is the greatest influence on the carbon cycle.'

To what extent do you agree with this view? **(20 marks)**

Total marks: 36

(Mark scheme and example answers on page 90)

Set 3

Question 9 The system concept and the carbon cycle

Explain the term 'natural system' as it relates to the carbon cycle. (4 marks)

(Mark scheme and example answers on page 93)

Question 10 The carbon cycle in the Arctic

Using **Figure 4** (see page 7) and your own knowledge, describe the possible changes to the carbon cycle in the Arctic and the impact of those changes. **(6 marks)**

(Mark scheme and example answers on page 94)

Question 11 Implications for the water cycle

Using **Figure 5** (see page 7) and your own knowledge, assess the implications of deforestation in the Amazon Basin for the water cycle. **(6 marks)**

(Mark scheme and example answers on page 95)

Question 12 Mitigating the impacts of climate change

'Human intervention is now the only way to mitigate the impacts of climate change.'

Using examples, assess the extent to which you agree with this statement. **(20 marks)**

Total marks: 36

(Mark scheme and example answers on page 97)

Section B option: Coastal systems and landscapes

Set 1

Question 1 Dynamic equilibrium

Explain how the concept of dynamic equilibrium can be applied to beach development. **(4 marks)**

(Mark scheme and example answers on page 100)

Question 2 Cliff processes

Using **Figure 6** (see page 8) and your own knowledge, explain the weathering processes at work on this cliff face. **(6 marks)**

(Mark scheme and example answers on page 101)

Question 3 Storm beaches and sediment

Table 1 shows data collected as part of a study on storm beaches and sediment. Complete the calculations in Table 1 and interpret your Spearman's rank test result using **Figure 7** (see page 8). **(6 marks)**

Null hypothesis: There is no relationship between the height of the storm beach and the size of the sediment.

Table 1 The relationship between the height of the storm beach and the size of the sediment ($n = 8$)

Height of storm beach (m)	Shingle diameter (cm)	R_1	R_2	d	d^2
5.5	8.4	1	8	–7	49
8	7.7	2	7	–5	25
11	6.1	5	3	2	4
10	6.2	3.5	4		0.25
10	7.5	3.5	6	–2.5	6.25
14	6.5	6	5	1	1
18	4.8	8	1	7	49
15	5.8	7	2	5	25

$\Sigma d^2 =$ ____________________

Formula for calculating Spearman's rank correlation coefficient:

$$R_s = 1 - \left(\frac{6 \Sigma d^2}{n^3 - n} \right)$$

$R_s =$ ____________________

....

....

....

....

(Mark scheme and example answers on page 103)

Question 4 Coastal management

'Sustainable solutions to coastal management require a variety of approaches.'

To what extent do you agree with this statement? **(20 marks)**

....

....

Total marks: 36

(Mark scheme and example answers on page 104)

Set 2

Question 5 High-energy and low-energy coasts

Explain the terms 'low-energy coasts' and 'high-energy coasts'. **(4 marks)**

(Mark scheme and example answers on page 108)

Question 6 Coastal stacks

Using **Figure 8** (see page 8) and your own knowledge, explain the processes that would have created this landscape feature. **(6 marks)**

(Mark scheme and example answers on page 109)

Question 7 GIS and shoreline management plans

Explain how a geographic information system (GIS) would help inform a local shoreline management plan (SMP). **(6 marks)**

(Mark scheme and example answers on page 110)

Question 8 Effect of sea-level changes

Assess the importance of the role played by changing sea levels in the formation of differing coastal landscape types. **(20 marks)**

Total marks: 36

(Mark scheme and example answers on page 112)

Section C option: Hazards

Set 1

Question 1 Tropical cyclones

Using **Figure 9** (see page 9) and your own knowledge, explain the conditions that can lead to the formation of a tropical cyclone. **(6 marks)**

(Mark scheme and example answers on page 115)

Question 2 Effects of storms

Using **Figures 10 and 11** (see page 9) and your own knowledge, assess the extent to which the effects of a storm in a country with a highly developed economy will be different from the effects in one that has an emerging economy. **(9 marks)**

(Mark scheme and example answers on page 117)

Question 3 Volcanic events

With reference to one or more volcanic events that you have studied, assess the extent to which the hazards created can be effectively managed. **(9 marks)**

(Mark scheme and example answers on page 120)

Question 4 A multi-hazardous environment

'Some countries or regions of the world are more prone to multiple natural hazards than others.'

To what extent do you agree with this view? **(20 marks)**

Total marks: 44

(Mark scheme and example answers on page 122)

Set 2

Question 5 Plate boundaries

Using **Figure 12** (see page 10) and your own knowledge, explain the processes at work at this type of plate boundary. **(6 marks)**

(Mark scheme and example answers on page 126)

Question 6 Wildfires

Using **Figure 13** (see page 10), analyse the data shown in the two graphs and assess the extent to which measures can be taken to deal with wildfires. **(9 marks)**

(Mark scheme and example answers on page 127)

Question 7 Managing hazard risk

For a hazard-prone area you have studied, how far do you agree with the statement that: 'A management plan is the answer to dealing with hazard risk'? **(9 marks)**

(Mark scheme and example answers on page 129)

Question 8 Tectonic processes

'Tectonic processes produce a wide variety of differing landscapes.'

To what extent do you agree with this view? **(20 marks)**

Total marks: 44

(Mark scheme and example answers on page 132)

Paper 2 Human geography

Resources

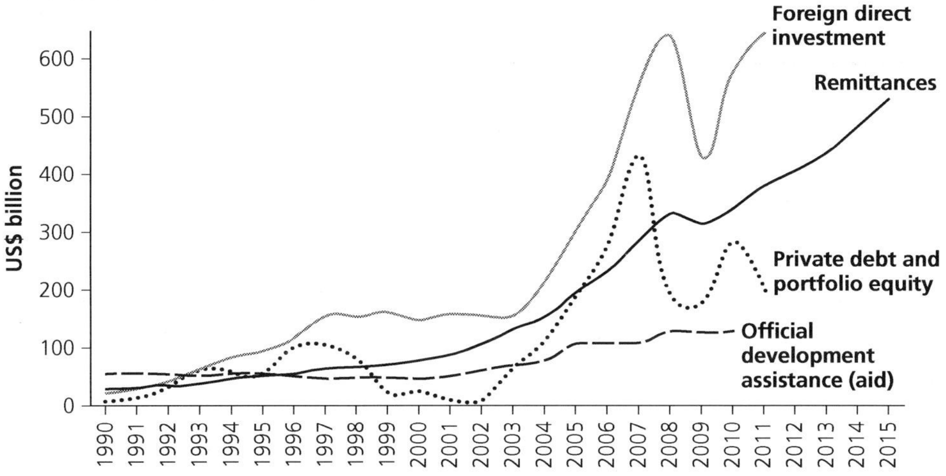

Figure 1 Flows of capital between richer areas and emerging markets (Source: data from World Development Indicators and World Bank estimates)

Year	Jan	Feb	Mar	Apr	May	Jun	Jul	Aug	Sep	Oct	Nov	Dec
2008	6.76	3.77	4.95	7.98	10.75	13.91	15.93	17.85	18.17	18.07	16.54	12.31
2018	4.24	2.26	3.19	5.84	8.99	12.67	15.54	17.13	17.63	17.91	15.38	9.96

Figure 2 The extent of Antarctic sea ice (millions of km2) (Source: National Snow and Ice Data Center, Boulder, CO)

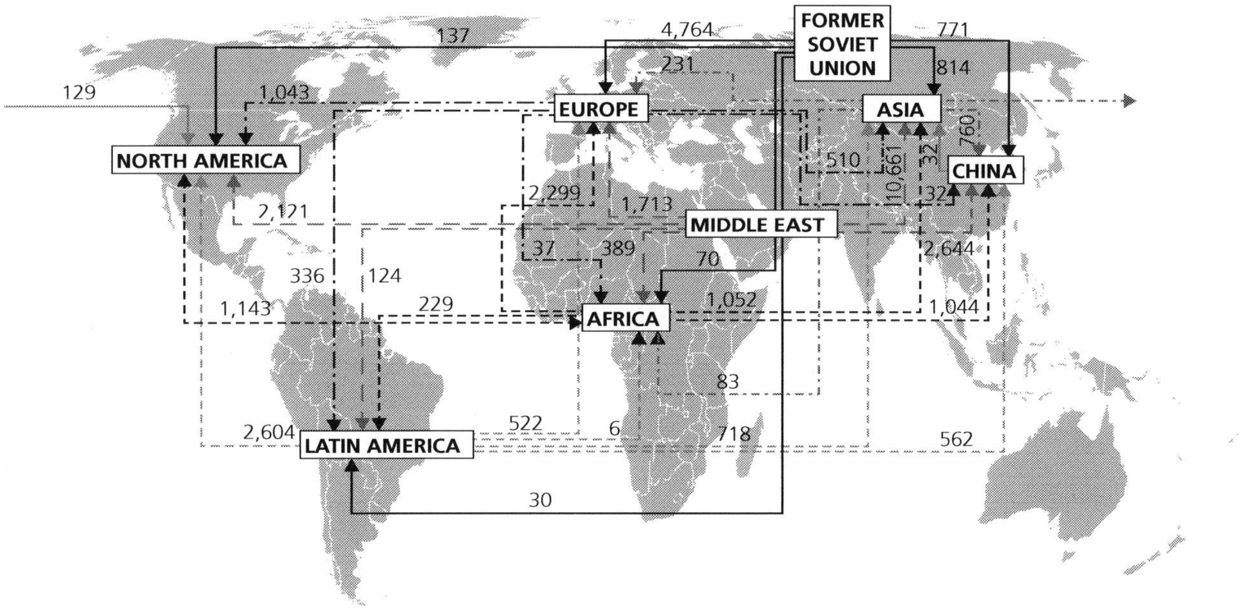

Figure 3 World crude oil exports by destination, 2013 (1000 barrels per day) (Source: OPEC Annual Statistical Bulletin, 2014)

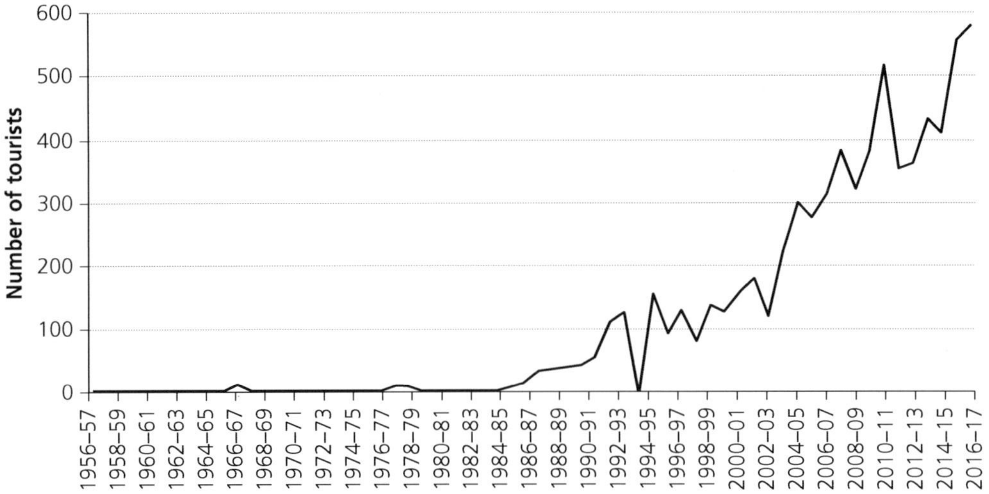

Figure 4 Land-based tourist arrivals in Antarctica (Source: IAATO)

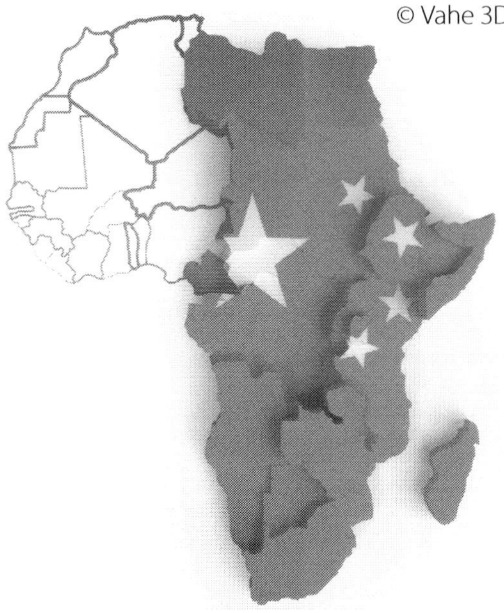

Figure 5 China and Africa

Who manages Antarctica?

No single country owns Antarctica. Instead, countries wishing to have a say in how the Antarctic (both the continent itself and the surrounding Southern Ocean) is governed must sign on to, and agree to abide by, the Antarctic Treaty.

Prior to the signing of the Antarctic Treaty in 1959, several countries had made claims to parts of Antarctica, some of which overlapped, and there are still disputes over territory that remain unresolved and come up from time to time.

The original signatories of the Treaty were Argentina, Australia, Belgium, Chile, France, Japan, New Zealand, Norway, South Africa, the UK the USA and the USSR. Since then, many other countries have acceded to the Antarctic Treaty.

The original Antarctic Treaty contains 14 articles, 11 of which relate directly to Antarctic governance. The other articles establish Antarctica as an area free from military activity and encourage scientific research and the exchange of scientific information between parties, as well as cooperation and peaceful conflict resolution.

The most important addition to the Treaty was the Madrid Protocol, formally known as the Protocol on Environmental Protection to the Antarctic Treaty, which was agreed in 1991 and came into force in 1998. The Protocol proclaims that protection of the Antarctic environment shall be considered paramount. It prohibits mineral and gas development and prevents unwise practices such as the dumping of waste from research bases directly into the Southern Ocean.

There are three other major agreements that are considered part of the Antarctic Treaty System. These protect various species such as seals, fish and crustaceans, as well as any plant life.

Among other agreements, there is the International Whaling Commission that is designed to ensure that whale stocks are conserved, as well as the Lindblad Plan and the International Association of Antarctica Tour Operators, the primary goal of which is to 'advocate and promote the practice of safe and environmentally responsible private-sector travel to the Antarctic'.

Figure 6 Article about who manages Antarctica

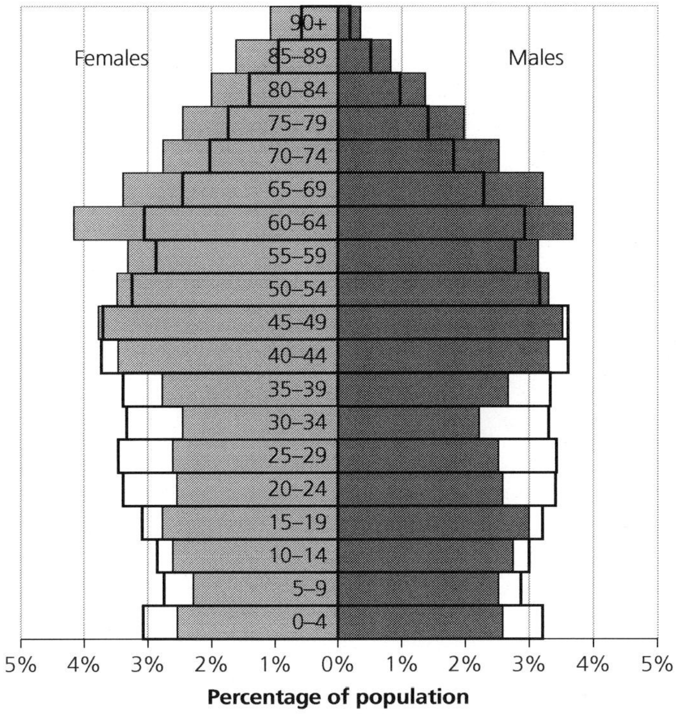

Figure 7 Population pyramid for Torbay, 2011 — the solid bars represent Torbay's resident population and the hollow bars represent the population structure in England

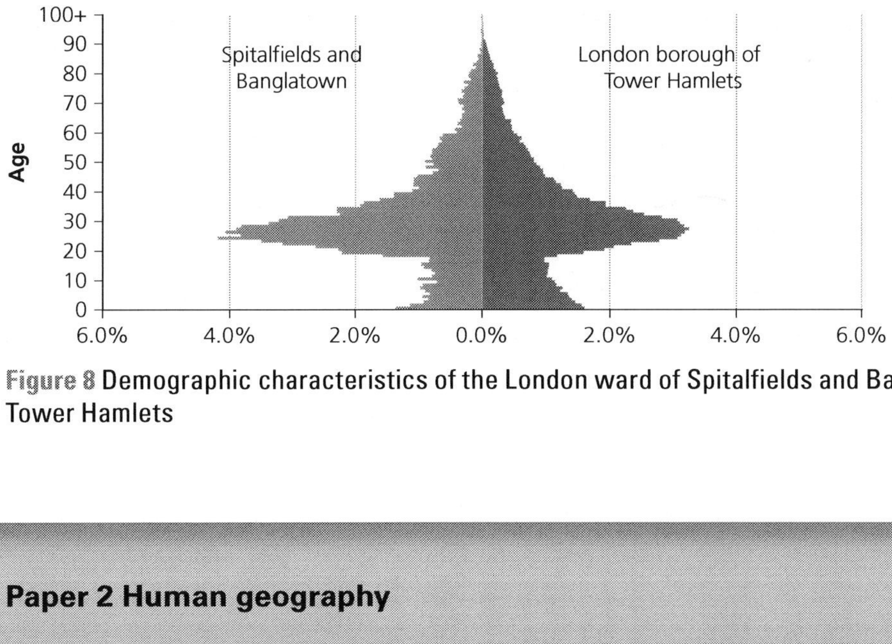

Figure 8 Demographic characteristics of the London ward of Spitalfields and Banglatown and the borough of Tower Hamlets

Fifteen people were asked to score ten environmental characteristics on a scale of 1 to 10. The scores were then added to give an overall environmental quality score that could range from 0 (a score of 0 for each of the ten environmental characteristics) to 100 (a score of 10 for each environmental characteristic).

The total scores for each of the 15 respondents are shown in the table below, with 0 indicating poor environmental quality and 100 excellent environmental quality.

Respondent	1	2	3	4	5	6	7	8	9	10	11	12	13	14	15
Urban setting: total environmental quality score	10	12	11	15	25	29	14	12	27	31	40	42	33	20	14
Rural setting: total environmental quality score	70	40	29	88	15	92	50	53	48	32	20	80	71	64	20

Figure 9 Data on the environmental perception of an urban setting and a rural setting

Figure 10 Photograph of Kettlewell in Upper Wharfedale, North Yorkshire

Figure 11 Photograph of Grassington Hub in Upper Wharfedale, North Yorkshire

Figure 12 An Eastern European shop

Figure 13 An Indian restaurant

Area of London	Number of boroughs with high levels of deprivation
Northeast	8
Northwest	7
Southwest	2
Southeast	3

Figure 14 Distribution of high levels of deprivation in London

Degrees of freedom	Significance level	
	0.05	**0.01**
3	7.82	11.34
4	9.49	13.28
5	11.08	15.09

Figure 15 Critical values for chi-squared

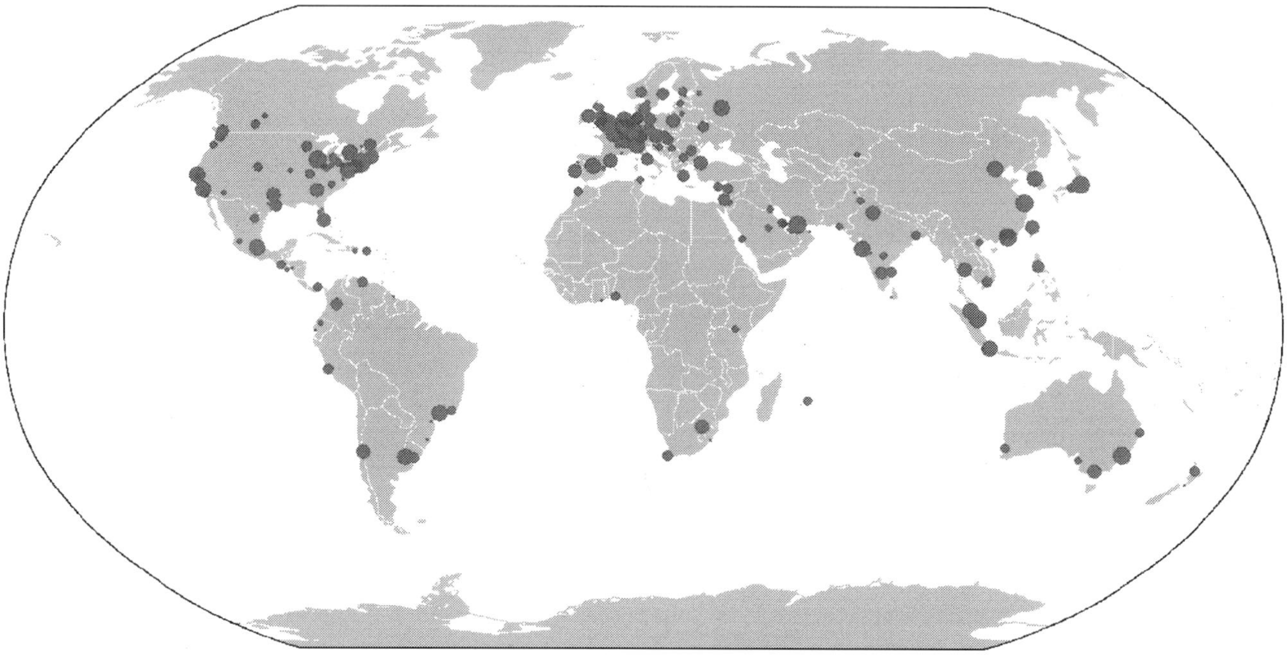

Figure 16 Map showing the distribution of GaWC-ranked world cities, 2010. GaWC is the commonly used abbreviation for the Globalization and World Cities Research Network based at Loughborough University (Source: https://commons.wikimedia.org/wiki/File:GaWC_World_Cities.png)

Figure 17 A city in the developing world

Distance from CBD (km)	0	2	4	6	8	10	12	14	16	18	20
Temperature (°C)	33.5	33	32	28	28.5	30	29	28	28	26	25

Figure 18 Temperature readings taken at data collection points with increasing distance from a city centre (CBD) on a summer day with light winds

Degrees of freedom	Significance level	
	0.05	**0.01**
10	0.564	0.745
11	0.523	0.736
12	0.497	0.703

Figure 19 Critical values for R_s

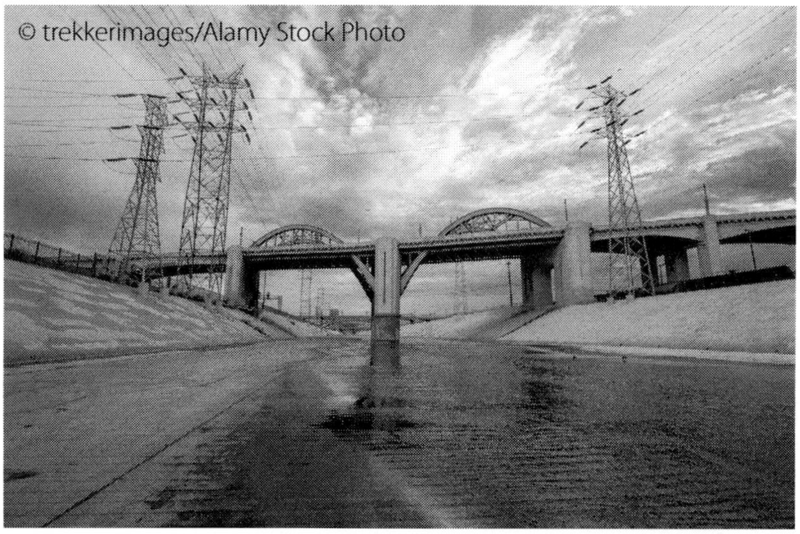

Figure 20 Photograph of the Los Angeles River

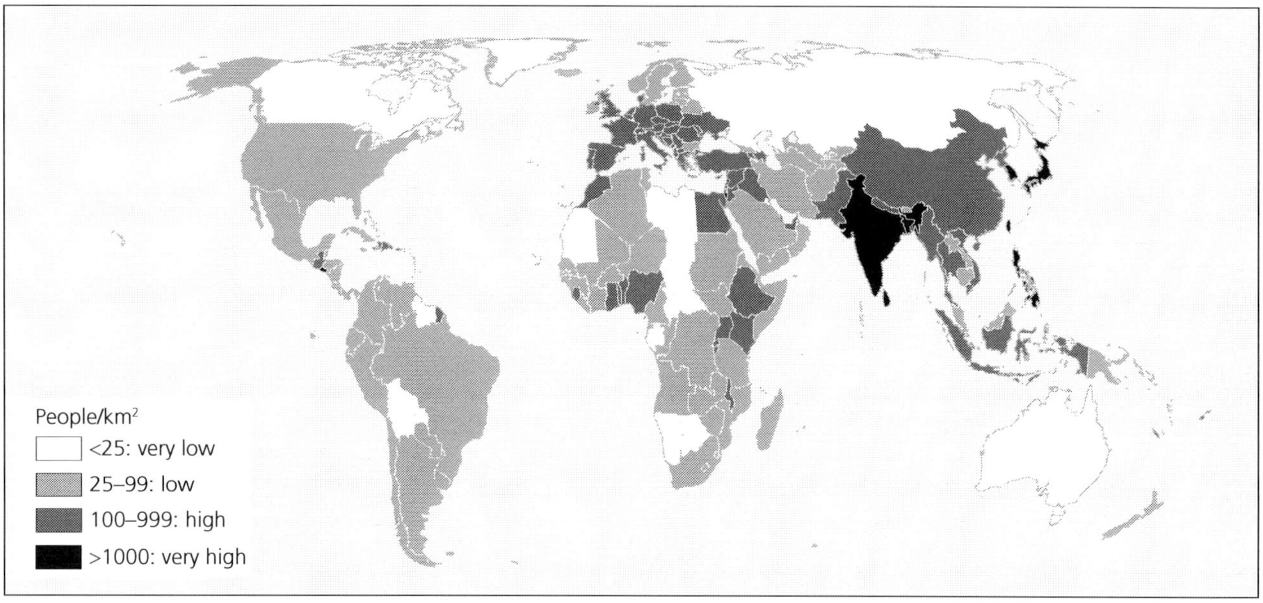

Figure 21 Population density, 2015

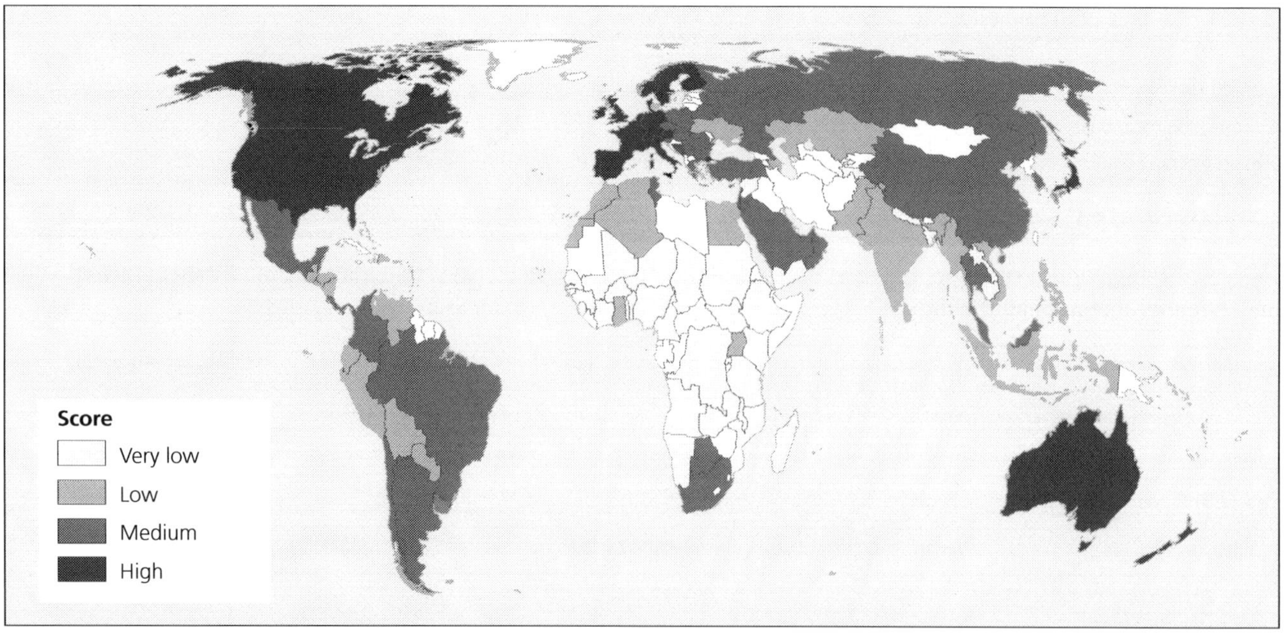

Figure 22 Global food security, 2015

Low-income countries		High-income countries	
Cause	**Deaths per 100,000**	**Cause**	**Deaths per 100,000**
Lower respiratory infections (e.g. pneumonia)	91	Ischaemic heart disease	158
HIV/AIDS	65	Stroke	95
Diarrhoeal diseases	53	Cancer (trachea/lung)	49
Stroke	52	Alzheimer's disease	42
Ischaemic heart disease	39	Chronic obstructive pulmonary disease (COPD)	31
Malaria	35	Lower respiratory infections	31

Figure 23 Differences in the top six causes of death between low- and high-income countries (deaths in rank order, per 100,000 population) (Source: World Health Organization)

Stage	1 High stationary High fluctuating	2 Early expanding	3 Late expanding	4 Low stationary Low fluctuating	5 Decline

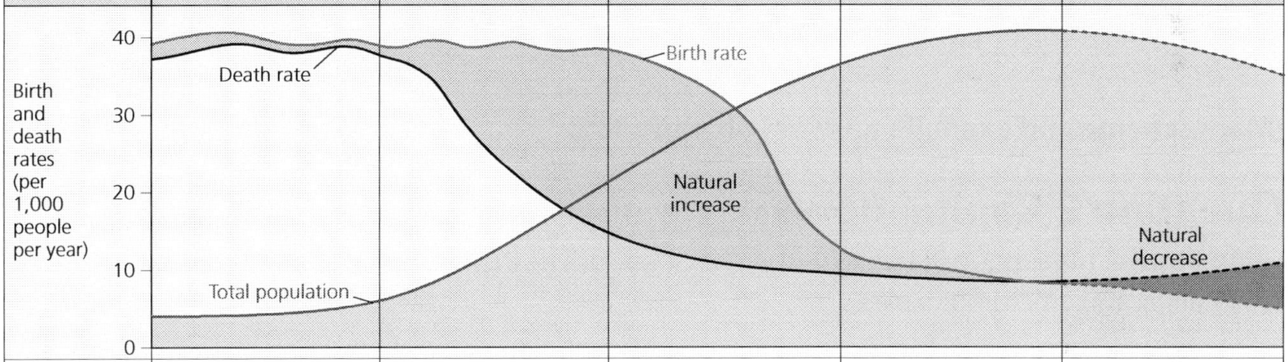

Examples	A few remote groups	Niger, Mali, Gambia	Brazil, India, Mexico	USA, Canada, UK	Germany, Japan, Russia
Birth rate	High	Remains high	Starts to fall	Low and steadies	Falls below DR
Death rate	High	Falls rapidly	Falls more slowly	Starts to level out	Remains low; may increase
Natural increase	Stable and fluctuating	Very rapid increase	Increase slows down	Slow increase; some fluctuations	Small natural decrease
Reasons for change in birth rate	Many children needed to work on the land. Many children die at an early age. Religious and social reasons. No family planning		Conditions change so fewer children needed. Lower IMR. Contraception	More family planning. Improved status of women with careers	More elderly population. Smaller proportion in reproductive ages
Reasons for change in death rate	Disease, famine. Poor medical knowledge so many children die	Improvements in food and clean water supply, sanitation and hygiene. Medical care advances, more vaccines. Fewer children die		Good healthcare established. Reliable food supply	More elderly population, higher proportion reaching life expectancy.

Figure 24 The demographic transition model

Section A Global systems and global governance

Set 1

Question 1 Globalisation and capital flows

Explain the impact of globalisation on flows of capital. **(4 marks)**

(Mark scheme and example answers on page 135)

Question 2 Capital flows

Using **Figure 1** (see page 35), analyse the characteristics of the flows of capital between richer areas and emerging markets. **(6 marks)**

(Mark scheme and example answers on page 136)

Question 3 Climate change and Antarctica

Using **Figure 2** (see page 35) and your own knowledge, assess the extent to which ice in the Antarctic is shrinking. **(6 marks)**

(Mark scheme and example answers on page 137)

Question 4 Transnational corporations (TNCs)

To what extent do transnational corporations (TNCs) bring more benefits than problems to the countries in which they operate? **(20 marks)**

Total marks: 36

(Mark scheme and example answers on page 138)

Set 2

Question 5 Global financial systems

Outline the role of any international organisation in supporting the global financial system. **(4 marks)**

(Mark scheme and example answers on page 142)

Question 6 Global flows of crude oil

Analyse the pattern of the global flows of crude oil as shown in **Figure 3** (see page 35). **(6 marks)**

(Mark scheme and example answers on page 143)

Question 7 Environment of Antarctica

Using **Figure 4** (see page 36) and your own knowledge, assess the potential impacts of the changes in tourist numbers on the Antarctic environment. **(6 marks)**

(Mark scheme and example answers on page 144)

Question 8 Impact of globalisation

To what extent has the impact of globalisation been uneven and/or unfair? **(20 marks)**

Total marks: 36

(Mark scheme and example answers on page 146)

Set 3

Question 9 Global commons

Explain the concept of the 'global commons'. **(4 marks)**

(Mark scheme and example answers on page 150)

Question 10 Inequalities between countries

Using **Figure 5** (see page 36), explain how some states drive global systems to their own advantage while others are only able to respond. **(6 marks)**

(Mark scheme and example answers on page 151)

Question 11 International government organisations and Antarctica

Using **Figure 6** (see pages 36–37), to what extent do you think the Antarctic is being well managed? **(6 marks)**

(Mark scheme and example answers on page 153)

Question 12 The work of the United Nations

'Since 1945 the United Nations has worked to promote growth and stability but has also exacerbated inequalities and injustices.'

To what extent do you agree with this statement? **(20 marks)**

Total marks: 36

(Mark scheme and example answers on page 154)

Section B Changing places

Set 1

Question 1 Aspects of place

Outline how any **two** endogenous factors contribute to the character of a place. **(4 marks)**

(Mark scheme and example answers on page 158)

Question 2 Population data for a place

Figures 7 and 8 (see page 37) show quantitative data that can be used to compare different places. Analyse the population data in Figures 7 and 8. **(6 marks)**

(Mark scheme and example answers on page 159)

Question 3 Place studies

Outline the strengths and weaknesses of two suitable sources of data when carrying out a place study. **(6 marks)**

(Mark scheme and example answers on page 160)

Question 4 Place meanings

With reference to one or more places you have studied, to what extent have attempts by external agencies to create specific place-meanings been successful? **(20 marks)**

Total marks: 36

(Mark scheme and example answers on page 162)

Set 2

Question 5 Cultural characteristics

Explain how the cultural characteristics of a place are shaped by shifting flows of people. **(4 marks)**

(Mark scheme and example answers on page 165)

Question 6 Perceptions of a place

Figure 9 (see page 38) shows data collected as part of a study on the perception of place carried out in an urban and in a rural setting. Evaluate the usefulness of the data in making comparisons of place perception in an urban and a rural area. **(6 marks)**

(Mark scheme and example answers on page 166)

Question 7 Character of places

Figures 10 and 11 (see pages 38 and 39) show two places in the Upper Wharfedale Valley of North Yorkshire.

Using Figures 10 and 11 and your own knowledge, suggest the types of factors that contribute to the character of these places. **(6 marks)**

(Mark scheme and example answers on page 167)

Question 8 Past and present connections

To what extent have past and present connections shaped one or more place(s) that you have studied? **(20 marks)**

Total marks: 36

(Mark scheme and example answers on page 169)

Set 3

Question 9 Representing a place

In what variety of ways might a place be represented? **(4 marks)**

(Mark scheme and example answers on page 172)

Question 10 Cultural characteristics of a place

How useful are **Figures 12 and 13** (see page 39) in helping to understand how cultural characteristics of a place are changed by flows of people? **(6 marks)**

(Mark scheme and example answers on page 173)

Question 11 Deprivation in London

Using **Figure 14** (see page 40), which shows data collected on the distribution of high levels of deprivation in London, complete the calculations in Table 1 and interpret your chi-squared result using **Figure 15** (page 40). **(6 marks)**

Null hypothesis: There is no significant difference in the distribution of high levels of deprivation in London.

Formula for calculating chi-squared:

$$\chi^2 = \Sigma \frac{(O - E)^2}{E}$$

Table 1 Chi-squared calculation

Area of London	Observed (O)	Expected (E)	($O - E$)	$(O - E)^2$	$\frac{(O-E)^2}{E}$
Northeast	8	5	3		1.8
Northwest	7	5	2	4	
Southwest	2	5	–3	9	1.8
Southeast	3	5	–2	4	0.8
Total	20	20	—	—	=

....

....

....

....

....

....

(Mark scheme and example answers on page 176)

Question 12 Lived experience of a place

'People's lived experience of place will depend on continuity and change in the nature of that place.'

With reference to one or more places you have studied, assess the validity of this statement.

(20 marks)

Total marks: 36

(Mark scheme and example answers on page 177)

Section C option: Contemporary urban environments

Set 1

Question 1 Distribution of world cities

Analyse the distribution of world cities shown in **Figure 16** (see page 40). **(6 marks)**

(Mark scheme and example answers on page 181)

Question 2 Economic inequality in urban areas

Using **Figure 17** (see page 41) and your own knowledge, analyse the issues associated with economic inequality in urban areas. **(9 marks)**

(Mark scheme and example answers on page 182)

Question 3 Waste management solutions

Assess the effectiveness of incineration and landfill as waste management solutions. **(9 marks)**

(Mark scheme and example answers on page 184)

Question 4 Sustainable urban areas

To what extent can effective planning and management create more sustainable urban areas? **(20 marks)**

Total marks: 44

(Mark scheme and example answers on page 187)

Set 2

Question 5 Urban climate

Figure 18 (see page 41) shows data collected as part of a study of urban climate. Complete the calculations in Table 1 and interpret your Spearman's rank test result using **Figure 19** (see page 41). **(6 marks)**

Null hypothesis: There is no significant relationship between distance from CBD and temperature.

Table 1 Calculating Spearman's rank correlation coefficient (R_s)

Data collection point	Distance from CBD (km)	Rank	Temperature (°C)	Rank	*d*	d^2
1	0	11	33.5	1	10	100
2	2	10	33	2	8	64
3	4	9	32	3	6	36
4	6	8	28	8	0	0
5	8	7	28.5	6	1	1
6	10	6	30	4	2	4
7	12	5	29	5	0	0
8	14	4	28	8		16
9	16	3	28	8	–5	25
10	18	2	26	10	–8	64
11	20	1	25	11	–10	100
						Total = 410

Formula for calculating the Spearman's rank correlation coefficient:

$$R_s = 1 - \left(\frac{6\sum d^2}{n^3 - n}\right) \qquad R_s = 1 - \left(\frac{2460}{n^3 - n}\right)$$

....

....

....

....

....

....

(Mark scheme and example answers on page 191)

Question 6 River catchments in urban areas

Using **Figure 20** (see page 42) and your own knowledge, assess the extent to which the management of river catchments in urban areas presents a particular challenge. **(9 marks)**

(Mark scheme and example answers on page 192)

Question 7 Urban deprivation

Assess the difficulties of managing urban deprivation. **(9 marks)**

(Mark scheme and example answers on page 195)

Question 8 Urban environmental issues

'Environmental issues pose the greatest challenge to urban managers.'

To what extent do you agree with this statement? **(20 marks)**

Total marks: 44

(Mark scheme and example answers on page 197)

Section C option: Population and the environment

Set 1

Question 1 Food security

Using **Figures 21 and 22** (see page 42), analyse the relationship between population density and food security. **(6 marks)**

(Mark scheme and example answers on page 201)

Question 2 Soil deterioration

Assess the extent to which poor agricultural practices lead to the deterioration of soils. **(9 marks)**

(Mark scheme and example answers on page 202)

Question 3 Overpopulation

Analyse the extent of the challenges faced by a country that is overpopulated. **(9 marks)**

(Mark scheme and example answers on page 205)

Question 4 International migration

'Countries, whatever their stage of development, gain economic benefits from international migration.'

To what extent do you agree with this statement? **(20 marks)**

Total marks: 44

(Mark scheme and example answers on page 208)

Set 2

Question 5 Mortality rates

Analyse the trends in **Figure 23** (see page 43), which shows the top six causes of death in high- and low-income countries. **(6 marks)**

(Mark scheme and example answers on page 213)

Question 6 Health and the physical environment

Analyse the relationship between health and the physical environment in **one** local area you have studied. **(9 marks)**

(Mark scheme and example answers on page 214)

Question 7 The demographic transition model

Using **Figure 24** (see page 43), to what extent can the demographic transition model be used to explain current trends in the population changes of individual countries? **(9 marks)**

(Mark scheme and example answers on page 216)

Question 8 Climate change and health

'Global environmental change is the biggest health threat of the twenty-first century — the impacts of which will be felt around the world.'

To what extent do you agree with this statement? **(20 marks)**

Total marks: 44

(Mark scheme and example answers on page 219)

MARK SCHEMES AND EXAMPLE RESPONSES

Paper 1 Physical geography

Section A Water and carbon cycles

Set 1

Question 1 mark scheme

4 marks (AO1 = 4 marks). 1 mark for each valid point, with marks available for development.

Notes for answers

- Overall view of flow: the movement of water between two stores within a drainage basin.
- Stem flow: when the precipitation is intercepted by vegetation; it then reaches the ground by flowing down stems, branches and trunks.
- Overland flow: when infiltration rates are exceeded or when the soil saturation point is reached.
- Throughflow: the movement of water under the influence of gravity through the subsoil.
- Base flow: the slowest of the flows — the movement of water in the saturated layer of rock.

Hints and tips

This question provides an opportunity to display extensive use of technical language.

Question 1 example responses

Student A

Water can flow through a drainage basin in several different ways. When precipitation occurs, vegetation will intercept it before it reaches the ground. The rain will then flow along branches and down the trunks of trees before becoming overland flow. Overland flow occurs when the precipitation is greater than the infiltration rate or when the soil is already saturated from a previous event. Once infiltrated into the soil, the water soaks through the soil pores or through naturally occurring pipes under gravity, down to the saturated layer. This is the zone below the water table. Base flow is the slowest of the flows and occurs in the saturated layer; it gets slower the deeper it gets below the surface.

e **The student identifies four major flows in the 'right' order of occurrence, and gives considerable detail. There is good use of technical language and a clear understanding of process. 4 marks**

Student B

There are a number of different ways that water can flow through a drainage basin: throughflow, overland flow and stem flow. Overland flow is where the water flows on the surface of the land, stem flow is when it runs down a tree and base flow is when it flows through rocks very slowly.

e **The student has not identified all flows and having named three has written about a different three. There is only superficial detail given on the flows. 3 marks**

Question 2 mark scheme

6 marks (AO1 = 2 marks, AO2 = 4 marks). Use the levels of response mark scheme as follows.

Level 2 (4–6 marks)

- **AO1** — Demonstrates clear knowledge and understanding of concepts, processes, interactions and change.
- **AO2** — Applies knowledge and understanding to the novel situation, offering clear analysis and evaluation drawn appropriately from the context provided. Connections and relationships between different aspects of study are evident, with clear relevance.

Level 1 (1–3 marks)

- **AO1** — Demonstrates basic knowledge and understanding of concepts, processes, interactions and change.
- **AO2** — Applies limited knowledge and understanding to the novel situation, offering basic analysis and evaluation drawn from the context provided. Connections and relationships between different aspects of study are basic, with limited relevance.

Notes for answers

AO1

- Potential changes to the climate may include: wetter winters; hotter, drier summers; increased frequency of storms.
- Uses made of groundwater include domestic, industrial and agricultural.

AO2

- Links made between changes in climate and changes to the levels of groundwater.
- Issues of recharge and problems as climate changes.
- Intrusion of sea water into aquifers near the coast or tidal estuaries.
- Loss of springs as groundwater levels sink.
- Regional differences, e.g. the southeast of the UK may be affected more severely due to much lower levels of precipitation and yet greater usage from growing population.
- Use of examples and precise use of technical language.

Hints and tips

This question requires you to use precise technical language and to give examples to illustrate your points. You must show that you have used Figure 1 in some way.

Question 2 example responses

Student A

Climate change in Britain is likely to bring more extreme weather conditions, with winters becoming wetter and stormier and summers becoming hotter and drier. This may be particularly true in the southeast, which is the area of the UK where groundwater is most important for water supply. Wetter winters might lead to an increase in groundwater, but water from storms tends to run to rivers so the increase might not be as great as expected. If, as predicted, the spring and autumn become drier and warmer, the period when groundwater levels in the aquifers might be recharged will be shorter so, overall, less water will be replaced.

During hot summers the main impact on groundwater will be the increase in demand and therefore higher extraction rates, especially if river flow is reduced, as was the case in Dorset in 2012. Higher rates of potential evapotranspiration may lead to farmers having to use irrigation methods such as on the coastal plain in West Sussex and in parts of East Anglia. Pumping water from deeper supplies in this way would threaten domestic sources. All of this may lead to a long-term decline in groundwater storage.

Another effect of climate change is likely to be rising sea levels. These may lead to the loss of groundwater resources in very low-lying coastal areas, as sea water can flow into the aquifers and contaminate the fresh water supply.

A general lowering of groundwater levels may well have an impact on vulnerable wetland ecosystems, which are important within the environment, and also lead to lower surface water levels of lakes and ponds that are used for recreation and tourism.

e The student uses good technical language and clearly understands the implications of climate change in the UK. The answer is long for 6 marks. Level 2

Student B

Groundwater supplies are very important for the water supply in the UK. Climate change is likely to bring hotter summers and wetter winters. If enough rain falls in the winter to supply the increased need in the summer there will not be a problem, but if we use more water in the summer than falls in the winter the reverse is true. Farmers, in particular, may find that their fields are flooded in the winter and parched in the summer. As the climate changes, the time available for recharge will get shorter and this will lead to a decline in groundwater resources. Coastal areas may also suffer as sea water comes in and fills the space taken by the groundwater that has been pumped out. Salt water cannot be used until it has been treated and it would mix with the fresh water, causing it to be unusable.

e The student has a less formal way of writing, and although the answer starts well it loses its way by referring to material that is outside the scope of this question. Some technical language is used, but overall there is a lack of detail. Level 1

Question 3 mark scheme

6 marks (AO1 = 2 marks, AO2 = 4 marks). Use the levels of response mark scheme as follows.

Level 2 (4–6 marks)

- **AO1** — Demonstrates clear knowledge and understanding of concepts, processes, interactions and change.
- **AO2** — Applies knowledge and understanding to the novel situation, offering clear analysis and evaluation drawn appropriately from the context provided. Connections and relationships between different aspects of study are evident, with clear relevance.

Level 1 (1–3 marks)

- **AO1** — Demonstrates basic knowledge and understanding of concepts, processes, interactions and change.
- **AO2** — Applies limited knowledge and understanding to the novel situation, offering basic analysis and evaluation drawn from the context provided. Connections and relationships between different aspects of study are basic, with limited relevance.

Notes for answers

AO1 Knowledge and understanding of the carbon cycle and the enhanced greenhouse effect and the factors that can lead to changes. For example:

- The carbon cycle is essentially nature's way of reusing carbon atoms in different ways and in varying places. It is the process by which carbon travels from the atmosphere into organisms and the Earth and then back into the atmosphere.
- The enhanced greenhouse effect is the process by which human activities — particularly burning fossil fuels (coal, oil and natural gas), agriculture and land clearing — are increasing the concentrations of greenhouse gases.

AO2 Application of knowledge and understanding to explain how changes to the cycle have led to an enhanced greenhouse effect. For example:

- Natural effects, volcanic activity and wildfires have a limited effect in the long term.
- Man-made or anthropogenic effects are believed by most scientists to have had a profound effect since the mid-nineteenth century.
- Changes can be to the terrestrial and the aquatic environment.
- Carbon emissions from burning hydrocarbons come from domestic, agricultural and industrial sources.
- Some industries create more carbon emissions than others, e.g. the power generation and cement industries.
- Deforestation is a major contributor to carbon emissions. Ploughing of land, which encourages microbial activity, and the removal of natural carbon sequestration sources through urban growth, all have significant effects.
- Changing sea temperatures can affect the oceanic carbon pump. For example, the increase in the temperature of the ocean weakens the ocean carbon sink. An increase of 2–3°C in sea surface temperature decreases the solubility of carbon dioxide by a few percent, and thus the capacity of the ocean to absorb carbon dioxide. This could further enhance atmospheric warming.

Hints and tips

This question centres on the connection between changes to the carbon cycle and changes to global climate, and your ability to explain that connection. You must show that you have used Figure 2 in some way.

Question 3 example responses

Student A

Changes to the carbon cycle can be seen as either natural changes or anthropogenic changes. While natural changes have taken place, their impact has not increased so much that it would have affected the natural greenhouse effect. Wildfires and volcanic activity continue to occur in much the same way that they always have. The major changes to the carbon cycle are those affected by humans. The first and most important of these is the burning of carbon-based fuels, which releases carbon dioxide into the air, enhancing the greenhouse effect. For example, in the cement industry, especially in countries like China, large amounts of carbon dioxide are released during the production process.

Deforestation and the burning of trees as in Amazonia releases carbon dioxide, and the deep ploughing of fields can release carbon from increased bacterial activity. Changes in land use from rural to urban not only increase carbon dioxide emissions but also reduce those landscapes that are capable of storing carbon as the land is cleared. The increase of carbon in the atmosphere allows it to absorb more re-radiated heat from the Earth and so causes the atmosphere to warm. This creates a positive feedback loop, which releases more carbon and so causes further heating.

e **The student gives a balanced account of the terrestrial changes to the carbon cycle that is more than adequate in the time available. Changes to the aquatic environment would have added to this answer. Good technical language is used. High Level 2**

Student B

Many changes have taken place in the carbon cycle, some natural and some man-made. Volcanic eruptions pour huge amounts of carbon dioxide into the air, affecting the greenhouse effect and causing temperatures to rise. Car fumes and fumes from central heating and power stations in towns can have the same effect, causing carbon dioxide to be released and creating acid rain. Cutting down trees in the Amazon means they cannot store carbon and instead it is put into the air when the trees are burned. All of this causes an enhanced greenhouse effect.

e **The student makes a basic attempt to address this question. Some valid points are made, but the answer is confused in places. The use of technical language is poor. High Level 1**

Question 4 mark scheme

20 marks (AO1 = 10 marks, AO2 = 10 marks). Use the levels of response mark scheme as follows.

Level 4 (16–20 marks)

AO1

- Detailed, relevant and appropriate knowledge and understanding of place(s) and environments.
- Full and accurate knowledge and understanding of key concepts.
- Detailed awareness of scale and temporal change where appropriate.

AO2

- Detailed, evaluative conclusion.
- Detailed application of knowledge and understanding throughout.
- Full evidence of links between knowledge and understanding.

Level 3 (11–15 marks)

AO1

- Generally clear and relevant knowledge and understanding of place(s) and environments.
- Generally clear and accurate knowledge and understanding of key concepts.
- Generally clear awareness of scale and temporal change where appropriate.

AO2

- Clear, evaluative conclusion.
- Generally clear application of knowledge and understanding.
- Generally clear evidence of links between knowledge and understanding.

Level 2 (6–10 marks)

AO1

- Some relevant knowledge and understanding of place(s) and environments.
- Some knowledge and understanding of key concepts.
- Some awareness of scale and temporal change where appropriate. There may be a few inaccuracies.

AO2

- Some sense of a conclusion.
- Some partially relevant application of knowledge and understanding.
- Some evidence of links between knowledge and understanding.

Level 1 (1–5 marks)

AO1

- Very limited knowledge and understanding of place(s) and environments.
- Very limited knowledge and understanding of key concepts.
- Very limited awareness of scale and temporal change where appropriate. There may be a number of inaccuracies.

AO2

- Very limited and/or unsupported conclusion.
- Very limited application of knowledge and understanding.
- Very limited evidence of links between knowledge and understanding.

Notes for answers

AO1 Knowledge and understanding of the role of vegetation. For example:

- Vegetation is a store of water as well as carbon.
- Changes in vegetation cover will affect these stores interdependently.
- Vegetation transfers water within the water cycle by root uptake and transpiration.
- Vegetation transfers carbon by absorption from the atmosphere during photosynthesis and releases it through decay.
- The availability of water influences the rates of photosynthesis, net primary production (NPP) and transpiration.

A02 Application of knowledge and understanding to evaluate the significance of the role of vegetation in linking the two cycles. For example:

- The significance of spatial and temporal scales.
- The disturbance of systems and the resultant positive and negative feedback.
- The carbon cycle could stop but the water cycle would continue.
- Do different types of vegetation make a difference?
- The relevance of other systems and stores in keeping the system going.
- The use of applied examples to illustrate the points made.

Hints and tips

Remember, to get to the higher levels you must 'assess the importance of vegetation' in linking the two cycles. One way is to state right at the beginning how important you think the role is. This allows you to use the rest of your answer to support your opening statement with links and facts.

The quality of your answer will be determined by the way in which your points are supported by reference to a place or places that you have studied. Depth, detail and up-to-date material will also benefit your answer.

Using paragraphs will help isolate your ideas and give your answer a much better structure. At the end of the answer you need to round things off with a brief summative statement that refers back to the question.

Question 4 example responses

Student A

To fully understand the significance of the role of vegetation in linking the water and carbon cycles, it is best to examine this in the context of an example, such as the Amazon Basin.

The Amazon Basin contains one of the largest and most diverse of all the rainforests and its huge area acts as a store for 20% of the world's carbon. The rate at which the forest has been growing is the result of the trees' ability to store carbon dioxide. From a growth peak in the late 1990s, however, there has been a steady decline in the ability of the forest to take up the increasing amounts of carbon dioxide. The problem is that for many years the rainforest has been deforested, leading to Brazil becoming the fourth largest polluter in the world. Increasing amounts of carbon dioxide have also led to trees growing faster and dying quicker, leading to considerable areas of die-back within the forests. Photosynthesis plays a large part in capturing carbon gas from the atmosphere and turning it into carbon as trees. This process would not take place without the presence of water in the plants, so the two cycles are linked very closely together. Without one, the other would not happen, although the water cycle could continue without the forest.

The area is also one of heavy rainfall, on average 200 mmpa but in some areas as high as 6000 mmpa. Much of this rain never reaches the ground as it is intercepted by the vegetation and evaporated back into the atmosphere. Huge amounts of water are taken up by the vegetation and so the forest acts as a massive store of water. Levels of transpiration are also very high, with water being put back into the atmosphere from the leaves of the trees. Transpired moisture contains salt particles that then act as condensation nuclei, assisting in rain-making. The loss of the forests and their transpiration contribution to the water cycle leads to a reduction in rainfall. Only about a third of the rainfall ever reaches the sea but it still accounts for some 15% of all fresh water entering the oceans.

Developments in the forest have affected both cycles. The deforestation exposes the ground, which leads to increased evaporation from the soil while at the same time exposing the thin soils to rapid erosion, making the ground unsuitable for vegetation. This evaporated moisture does not produce the typical rain-bearing clouds and so rainfall totals decrease, again making regrowth of the trees problematic.

The changes in vegetation from forest to pasture also affect the two cycles. Soil temperatures are higher on pasture land and as a result soil moisture content is lower. The tree root systems are able to release far more moisture into the air and therefore increase humidity levels and the likelihood of rain.

All these factors show how vegetation plays a vital link between the water and carbon cycles.

e The student provides detail and place specifics that relate directly to the question and explain how different processes link the two cycles. **Level 4**

Student B

The water cycle and the carbon cycle are both very important and work together. The plants that make up an important part of the carbon cycle need water to survive and so the whole thing is linked. The carbon cycle is the way that carbon is circulated through the environment, with plants able to absorb carbon dioxide and convert it into food to enable them to grow, but to do this the plants need water. The same applies to the oceans, where plants and animals all form part of the carbon cycle. The ocean currents move warm and cold water around, releasing and absorbing carbon dioxide as they go. On land, the Amazon Basin is a good example of how vegetation links the two cycles. The forests act as massive stores of carbon, taking the gas and converting it into trees themselves. The rainfall helps this process to take place. Without the forest, there would be no evapotranspiration and no moisture would be returned to the air by the plants. This is important as the moisture then is converted into rain that falls back on to the forest. There has been a lot of deforestation in Brazil and the trees have been burned, releasing carbon into the air, but there are no trees left to absorb it back, creating a greenhouse effect. Also, without the trees, the rainfall hits the ground and flows into the rivers, taking the soil with it and making the rainforest into a desert where no trees will grow. The lack of trees means that rainfall also goes down, making the problem worse.

It is very important for the planet that the cycles work together and to do that they need the vegetation to link everything up.

e The student attempts to answer the question and has some understanding of the issues. However, the example is not fully exploited and there are areas of confusion, so the response is rather disjointed. The use of paragraphs would have helped clarify the student's thoughts and therefore helped with the structure of the answer. **Level 2**

Set 2

Question 5 mark scheme

4 marks (AO1 = 4 marks). 1 mark for each valid point, with marks available for development.

Notes for answers

- Feedback is part of the dynamic equilibrium of the Earth and its systems.
- Positive feedback is where the effects of an action are increased by knock-on secondary effects. For example, more carbon dioxide in the atmosphere acts as a greenhouse gas and leads to a rise in global temperatures and warmer oceans; warm water is less able to dissolve the gas so carbon dioxide is released by the warmer oceans and more carbon dioxide is put back into the atmosphere.
- Negative feedback is where the effects of an action are negated by subsequent knock-on effects. For example, the increased use of fossil fuels causes a rise in global temperatures leading to global warming, which increases plant growth; more plants result in a greater take-up of carbon dioxide, reducing the amount in the atmosphere, and this reduction has the effect of reducing global temperatures.

Question 5 example responses

Student A

Positive feedback is where things react positively following an action. If there is more carbon dioxide in the atmosphere, this will lead to global warming, which in turn will warm the oceans. Warm water cannot hold as much carbon dioxide and so the gas is released back into the air, which means there is more carbon dioxide in the atmosphere. More carbon dioxide will cause more plant growth, which could then remove the gas, leading to a drop in temperature and a negative feedback — this is how dynamic equilibrium works.

e The student has a clear understanding of the concept and offers good examples. 4 marks

Student B

Negative feedback is when something happens and then it all goes wrong. If we burn a lot of fossil fuels, this leads to more carbon dioxide in the air, which leads to global warming and sea levels rising. This means people will have to move if their houses are in danger and they may have to burn more fossil fuels to get dry.

e The student has missed the point here, as the sequence will not end up balancing itself back into a positive cycle. 1 mark

Question 6 mark scheme

6 marks (AO3 = 6 marks)

Completing the calculation (3 marks)

1 mark for two correct completions within the table; 2 marks for all four. 1 mark for both correct calculations.

Year	Average atmospheric temperature compared with long-term average temperature (1900–2010) (°C)	Rank R_1	Average CO_2 measured at Mauna Loa Observatory, Hawaii (ppm)	Rank R_2	Difference in ranks (d)	d^2
1970	0	9.5	328	10	–0.5	0.25
1975	0	9.5	330	9	0.5	0.25
1980	0.3	7	338	8	–1	1
1985	0.1	8	346	7	1	1
1990	0.4	5.5	353	6	–0.5	0.25
1995	0.5	**4**	361	**5**	**–1**	**1**
2000	0.4	5.5	369	4	1.5	2.25
2005	0.7	2.5	379	3	–0.5	0.25
2010	0.7	2.5	390	2	0.5	0.25
2015	0.9	1	401	1	0	0

$\Sigma d^2 = 6.5$ $R_s = +0.961$

Interpretation (3 marks)

Conclusion: The null hypothesis is rejected.

Note: Allow 3 marks for a correct interpretation against an incorrect calculation of R_s.

Hints and tips

This question tests your ability to do simple calculations and comment geographically on the result.

Question 6 example responses

Student A

1995	0.5	**4**	361	**5**	**–1**	**1**

$\Sigma d^2 = 6.5$ $R_s = +0.96$

The calculation is above the critical value at both the 0.05 and 0.01 significance levels. This means that the null hypothesis can be rejected and it can be stated that there is a significant relationship between the amount of atmospheric carbon dioxide and rising temperatures.

e The student completes the table and calculations accurately and fully interprets the result.

3 + 3 = 6 marks

Student B

1995	0.5	**4**	361	**5**	**–1**	**1**

$\Sigma d^2 = 6.5$ $R_s = +0.85$

The null hypothesis can be rejected.

e The student completes the table accurately but gets the final result wrong. They partially interpret the result, but this is lacking in detail. 2 + 1 = 3 marks

Question 7 mark scheme

6 marks (AO1 = 2 marks, AO2 = 4 marks). Use the levels of response mark scheme as follows.

Level 2 (4–6 marks)

- **AO1** — Demonstrates clear knowledge and understanding of concepts, processes, interactions and change.
- **AO2** — Applies knowledge and understanding to the novel situation, offering clear analysis and evaluation drawn appropriately from the context provided. Connections and relationships between different aspects of study are evident, with clear relevance.

Level 1 (1–3 marks)

- **AO1** — Demonstrates basic knowledge and understanding of concepts, processes, interactions and change.
- **AO2** — Applies limited knowledge and understanding to the novel situation, offering basic analysis and evaluation drawn from the context provided. Connections and relationships between different aspects of study are basic, with limited relevance.

Notes for answers

AO1

- Knowledge and understanding include the four phases of the soil moisture budget: surplus, utilisation, deficit and recharge.
- Knowledge of the possible changes to the climate of southern England as temperatures rise, i.e. warmer (+3°C), wetter (+10–30%) winters, with warmer (+4°C), drier (–40%) summers.

AO2 Application of knowledge and understanding of climatic change to the soil moisture budget. For example:

- Wetter winters will recharge soil moisture and create a surplus, leading to flooding.
- Higher temperatures in summer will lead to greater utilisation as potential evapotranspiration (PET) increases, leading to deficit.
- Extraction may increase, leading to greater and longer deficits exacerbated by reduced recharge.

There will be shorter but intense periods of rain, which may not have time to soak in but will merely run off, so recharge may be restricted.

Hints and tips

You must show that you have made use of Figure 3. To access higher marks in Level 2 you must make clear the links between warming climate and changes to the soil moisture budget, with some indication of the importance of those links.

Question 7 example responses

Student A

The soil water budget relies on a regular regime of rainfall to maintain it. In southern England, winter rainfall recharges the soil water budget so that it can be used during the following year. Rivers can run dry when this regime is interrupted, as is shown in the photograph. In a warming climate, rainfall is projected to come in shorter but much heavier downpours. This type of rain has less chance of soaking in as it will quickly run off, which means that there will be less recharge taking place. The warm weather will also mean that potential evapotranspiration will be higher and so the deficit position may be reached sooner and last longer. It may mean that there is no surplus at the start of the year and the soil water budget could run at an increasing deficit, meaning that there is no water for the utilisation phase.

e The student has a clear understanding of the soil water budget and possible problems, refers to the resource and makes competent use of technical terms. High Level 2

Student B

The warming climate will have a big impact on the soil water budget. Plants will need more water and this will shorten the period of utilisation, meaning that the deficit situation will be reached sooner. If it does not rain as much, there will be less recharge and therefore no surplus at the beginning of the year.

e In this rather short answer, the student uses all the technical terms but does not address the issue of a warming climate in any detail. Middle Level 1

Question 8 mark scheme

20 marks (AO1 = 10 marks, AO2 = 10 marks). Use the levels of response mark scheme as follows.

Level 4 (16–20 marks)

AO1

- Detailed, relevant and appropriate knowledge and understanding of place(s) and environments.
- Full and accurate knowledge and understanding of key concepts.
- Detailed awareness of scale and temporal change where appropriate.

AO2

- Detailed, evaluative conclusion.
- Detailed application of knowledge and understanding throughout.
- Full evidence of links between knowledge and understanding.

Level 3 (11–15 marks)

AO1

- Generally clear and relevant knowledge and understanding of place(s) and environments.
- Generally clear and accurate knowledge and understanding of key concepts.
- Generally clear awareness of scale and temporal change where appropriate.

AO2

- Clear, evaluative conclusion.
- Generally clear application of knowledge and understanding.
- Generally clear evidence of links between knowledge and understanding.

Level 2 (6–10 marks)

AO1

- Some relevant knowledge and understanding of place(s) and environments.
- Some knowledge and understanding of key concepts.
- Some awareness of scale and temporal change where appropriate. There may be a few inaccuracies.

AO2

- Some sense of a conclusion.
- Some partially relevant application of knowledge and understanding.
- Some evidence of links between knowledge and understanding.

Level 1 (1–5 marks)

AO1

- Very limited knowledge and understanding of place(s) and environments.
- Very limited knowledge and understanding of key concepts.
- Very limited awareness of scale and temporal change where appropriate. There may be a number of inaccuracies.

AO2

- Very limited and/or unsupported conclusion.
- Very limited application of knowledge and understanding.
- Very limited evidence of links between knowledge and understanding.

Notes for answers

AO1 Knowledge and understanding of carbon emissions. For example:

- Growing demand leads to deforestation but also afforestation.
- Increased economic growth leads to increased carbon emissions.
- The changing temperatures in the oceans that result from global warming may affect their role in the carbon cycle.
- The development of biofuels may reduce carbon emissions.
- The increased use of renewables may reduce carbon emissions.

AO2 Application of knowledge and understanding of the effects of the demand for energy on the carbon cycle. For example:

- Economic growth remains a driving force for most governments despite the impacts on the environment and the social risks involved.
- Fossil fuels are the dominant drivers of the global economy.
- Oil is predominant in the transport industry, while coal still dominates the worldwide production of electricity.
- Fossil fuels and deforestation are the two major factors affecting the carbon cycle.
- LICs and MICs use proportionately more fossil fuels year on year as they strive for economic growth, suggesting no slackening of demand.

- Change of land use is related to changing agricultural practices but also includes deforestation and monoculture, e.g. palm oil, and intensified agriculture also has higher energy demands.
- Some fossil fuel extraction methods create land use change and environmental damage, e.g. tar sands, peat extraction, mountaintop removal and deforestation.

Hints and tips

The quality of your answer will be determined by the way in which your points are supported by reference to a place or places that you have studied. Depth, detail and up-to-date material will also benefit your answer.

Start the answer with a sentence saying how much you agree with the statement under discussion. Using paragraphs will then help to consolidate your ideas and will give your answer a much better structure. Other demands, such as agriculture, are acceptable as a counter-argument.

To get to the higher levels you must relate your conclusion to your opening statement and the quote in the question. Do not forget to finish with a summative statement that draws all your arguments together.

Question 8 example responses

Student A

The growth in the demand for energy has a great effect on the carbon cycle, from an increase in emissions to changes in the oceanic carbon cycle. The burning of fossil fuels is the most obvious cause of increased emissions. As LICs strive to improve their economies, their demand increases for fuel to power their growing transport and industrial needs, for example the production of cement. Economic growth is seen as the priority despite the environmental damage and the risks associated with such advances. An example of the impact on the carbon cycle is the urban air pollution in China — despite the government's pronouncements and efforts to solve the problem, massive amounts of carbon dioxide are still released into the air.

Oil and coal remain vital to any economy, with transport systems needing oil and an electricity generation industry needing coal. Again, China is a good example. Huge amounts of coal are still mined and burned to supply the country's industrial needs and more coal-fired power stations are built every year, despite the building of the Three Gorges Dam, which was supposed to help solve the problem of coal use. The extraction of many fuels also creates high levels of emissions — fracking, shale gas and tar sand oil extraction are hyped up as 'cheaper' fuels but environmentally are very costly.

Land use changes have also affected the carbon cycle. Deforestation and the burning of timber releases massive amounts of carbon back into the atmosphere. Deforestation occurs for several reasons, for example clearing the land prior to fossil fuel extraction, but also often for agriculture to feed growing populations in LICs and MICs such as Brazil. In HICs such as the UK and in some MICs such as Malaysia, the intensification of agriculture has required greater inputs of fuel for mechanisation, automation and the establishment of artificial growing environments. There has been land clearance for the growing of biofuels, but any carbon impact is offset by the fact that they produce fewer emissions when they are burnt. In most research, it is becoming clear that human activity is creating climate change by the burning of fossil fuels, which provides both positive and negative feedbacks to the carbon cycle.

The hope has to be that the growth of renewables will see a reduction in emissions and a reduction in the impact on the carbon cycle. However, at present, the demand for energy and its associated activities are having a major impact on the carbon cycle.

e This wide-ranging answer covers a variety of topics. Greater detail and an expansion of the examples would have ensured a top mark, but there is good use of technical terms. Level 4

Student B

Energy is the burning of fossil fuels to power industry, drive vehicles and make electricity. All countries do this, but LICs use more pro rata as they are developing their economies. Their need to join the 'world club' is their priority and so they are not bothered about any damage they might be doing to the environment. China is a prime example of a country that burns huge amounts of coal to produce electricity and therefore pumps tonnes of carbon dioxide into the air, adding to global warming and air pollution — but it does not care.

Land clearance is another way of affecting the carbon cycle. In LICs and MICs, land is cleared for agriculture as these countries need to feed their growing populations. Deforestation in the Amazon Basin has been done for this purpose and the carbon balance has been upset. In HICs, agriculture has been intensified and the need for fuel for machines has grown, so that farming is now producing more emissions than it used to. Intensive cattle farming also produces emissions, but this is methane not carbon.

ⓔ **The student produces a largely descriptive answer and does not offer much in the way of assessment. Some good points are introduced, but they are not expanded on and there is a lack of detail.** **Level 2**

Set 3

Question 9 mark scheme

4 marks (AO1 = 4 marks). 1 mark for each valid point, with marks available for development. Reserve 2 marks for clear links to the carbon cycle.

Notes for answers

- Isolated, open and closed systems.
- Inputs, flows, stores and outputs; positive and negative feedback.
- More carbon dioxide in the atmosphere causes global temperatures to rise.
- An increase in ocean temperatures causes the release of dissolved carbon dioxide.
- Increased use of fossil fuels leads to an increase in atmospheric carbon dioxide.
- This in turn encourages plant growth, which reduces atmospheric carbon dioxide.
- Reduced carbon dioxide reduces global temperatures.

Hints and tips

This question provides an opportunity to display an extensive use of technical language.

Question 9 example responses

Student A

There are three types of system — isolated, closed and open. The carbon cycle is an example of a closed system. This means there are flows of energy but not matter into and beyond the system boundary. Inputs into the system would include the burning of fossil fuels. There would then be a flow of carbon dioxide into the atmosphere, where it would be stored. If this then led to a rise in temperatures that in turn led to carbon dioxide being released from the oceans as a result of warming, this could be seen as a positive feedback loop. A negative feedback would be increased plant growth resulting in less carbon dioxide in the atmosphere, leading to a reduction in temperatures.

ⓔ **The student easily covers four points.** **4 marks**

Student B

The carbon cycle is a closed system, which means that although carbon can move around, it never leaves the system — like a central heating system in a house. Inputs might come from burning coal and the carbon would be stored in the atmosphere. There would be an output of dissolved carbon gas falling as acid rain. This would be a negative feedback, although you can have positive feedback as well.

e The student covers two points (there are some inaccuracies) — development or an additional point is needed. **2 marks**

Question 10 mark scheme

6 marks (AO1 = 2 marks, AO2 = 4 marks). Use the levels of response mark scheme as follows.

Level 2 (4–6 marks)

- **AO1** — Demonstrates clear knowledge and understanding of concepts, processes, interactions and change.
- **AO2** — Applies knowledge and understanding to the novel situation, offering clear analysis and evaluation drawn appropriately from the context provided. Connections and relationships between different aspects of study are evident, with clear relevance.

Level 1 (1–3 marks)

- **AO1** — Demonstrates basic knowledge and understanding of concepts, processes, interactions and change.
- **AO2** — Applies limited knowledge and understanding to the novel situation, offering basic analysis and evaluation drawn from the context provided. Connections and relationships between different aspects of study are basic, with limited relevance.

Notes for answers

- Melting of the sea ice is leading to a reduction of ice cover at the North Pole.
- Warmer water means less carbon will sink to the bottom, affecting the carbon pump.
- Warmer waters might decrease the abundance of phytoplankton, reducing the sea's ability to take carbon from the atmosphere.
- Retreating sea ice, up to 40% in the last 35 years, will affect the lower trophic levels of the food chain — no algae, no krill, no fish, no seals, no polar bears.
- Rapid thawing of land surfaces in summer months.
- Release of methane from permafrost could lead to positive feedback.
- Changes in habitat will affect land-based mammals.

Hints and tips

You must at some stage refer to Figure 4, although it is not the only point of reference. You will need to supplement the information in the resource with your own knowledge.

Question 10 example responses

Student A

The years since 1960 have seen a temperature rise of between 2 and 3 degrees centigrade in the Arctic regions. This may have already had, and will continue to have, dramatic effects on the carbon cycle. Areas of land in this region will thaw out quicker in the summer and may not refreeze, which could lead to a release of methane from the permafrost into the atmosphere. More importantly, the changes to the oceans will be significant. Warmer waters mean that less carbon dioxide will be taken deep into the oceans as the warmer waters will not sink as deep.

However, it is the loss of phytoplankton in the warmer waters that will reduce the ocean's ability to take carbon from the atmosphere. This issue of lower trophic levels in the food chain may well in turn have a catastrophic effect on those creatures in the higher trophic levels. These are examples of positive and negative feedback loops. Changes in land-based habitats will also affect land-based mammals.

e The student achieves Level 2 by showing clear understanding of the issues, with quoted figures. 5 marks

Student B

The Arctic has got a lot warmer, unlike some places that have actually got colder. If the ground thaws out methane gas will be released, which will help push temperatures even higher. If the water gets warmer the animals that live in cold water will die out and the food chains will become disrupted. This means that less carbon will be taken from the atmosphere.

e There is some basic description but the response lacks focus on the Arctic. There is no quotation from the resource. Further evidence and more connections are needed. 2 marks

Question 11 mark scheme

6 marks (AO1 = 2 marks, AO2 = 4 marks). Use the levels of response mark scheme as follows.

Level 2 (4–6 marks)

- **AO1** — Demonstrates clear knowledge and understanding of concepts, processes, interactions and change.
- **AO2** — Applies knowledge and understanding to the novel situation, offering clear analysis and evaluation drawn appropriately from the context provided. Connections and relationships between different aspects of study are evident, with clear relevance.

Level 1 (1–3 marks)

- **AO1** — Demonstrates basic knowledge and understanding of concepts, processes, interactions and change.
- **AO2** — Applies limited knowledge and understanding to the novel situation, offering basic analysis and evaluation drawn from the context provided. Connections and relationships between different aspects of study are basic, with limited relevance.

Notes for answers

AO1

- Deforestation reduces the amount of potential evapotranspiration (PET) — a crucial part of the water cycle.
- Water vapour is transported by north easterlies to São Paulo.
- Deforestation increases stream flow.
- Increased soil erosion and localised flooding.

AO2

- Some planting has taken place but covering only a small area, given the amount of overall felling.
- Deforestation means regrowth has smaller leaves and therefore needs less water.
- Implications of water shortages for urban areas.
- Amazon farmers are clearing forest, thereby reducing the supply of water they need for their crops.

Hints and tips

There must be at least one clear reference to the resource. The key to accessing the four AO2 marks is in the assessment. This involves setting out issues and reaching a judgement.

Question 11 example responses

Student A

Deforestation in the Amazon Basin has serious implications for the water cycle, both in the area and further afield. The trees play a vital part in the water cycle by taking up water from the ground and, through potential evapotranspiration (PET), returning it to the atmosphere. This water vapour is then transported to São Paulo, where it falls as rain to supply the city with its water needs. A reduction in the forest means a reduction in PET and so the 'biotic pump' will no longer function as effectively, if at all.

Deforestation can also result in increased soil erosion and localised flooding as a result of increased stream flow. Any new trees that might be planted (and some have been, but nowhere near enough) will be smaller and therefore will take up less water and produce less PET. All of this has implications for the cities that rely on the rainfall for their needs. In addition, the farmers who are clearing the forest to grow crops are reducing the water supply for these crops, creating a problem for themselves.

It is clear, as the article states, that far more must be done to preserve the forest in order to preserve the water cycle in this region.

e The student covers a range of content, showing good knowledge and understanding that is supported by examples. There is application of this knowledge. 6 marks

Student B

The article talks about the lack of rain and the impact of deforestation on São Paulo and other Brazilian cities. The trees take up water from the ground and breathe it out as water vapour. This is then blown as clouds to São Paulo, where cooling takes place and then it rains, supplying the city with its water. Because trees are being deforested, there is less water vapour to form clouds and so São Paulo faces a crisis. More trees have to be planted to replace the trees that have been cut down to prevent a crisis like they had in 1931.

e The student shows some limited knowledge and understanding but has misinterpreted the article. There is no use of technical language relating to this issue. 3 marks

Question 12 mark scheme

20 marks (AO1 = 10 marks, AO2 = 10 marks). Use the levels of response mark scheme as follows.

Level 4 (16–20 marks)

AO1

- Detailed, relevant and appropriate knowledge and understanding of place(s) and environments.
- Full and accurate knowledge and understanding of key concepts.
- Detailed awareness of scale and temporal change where appropriate.

AO2

- Detailed, evaluative conclusion.
- Detailed application of knowledge and understanding throughout.
- Full evidence of links between knowledge and understanding.

Level 3 (11–15 marks)

AO1

- Generally clear and relevant knowledge and understanding of place(s) and environments.
- Generally clear and accurate knowledge and understanding of key concepts.
- Generally clear awareness of scale and temporal change where appropriate.

AO2

- Clear, evaluative conclusion.
- Generally clear application of knowledge and understanding.
- Generally clear evidence of links between knowledge and understanding.

Level 2 (6–10 marks)

AO1

- Some relevant knowledge and understanding of place(s) and environments.
- Some knowledge and understanding of key concepts.
- Some awareness of scale and temporal change where appropriate. There may be a few inaccuracies.

A02

- Some sense of a conclusion.
- Some partially relevant application of knowledge and understanding.
- Some evidence of links between knowledge and understanding.

Level 1 (1–5 marks)

A01

- Very limited knowledge and understanding of place(s) and environments.
- Very limited knowledge and understanding of key concepts.
- Very limited awareness of scale and temporal change where appropriate. There may be a number of inaccuracies.

A02

- Very limited and/or unsupported conclusion.
- Very limited application of knowledge and understanding.
- Very limited evidence of links between knowledge and understanding.

Notes for answers

There needs to be a clear understanding of the term 'mitigation'. Detail and/or examples in breadth or depth — local, national or international.

A01

- Clear explanation/definition of mitigation and perhaps how it differs from adaptation.
- Use of technical language, such as carbon capture and sequestration (CCS), renewables.
- Knowledge of carbon capture, transport and storage.
- Knowledge of the technologies associated with these terms.

A02

- Understanding and detail of how improved aviation practices would help mitigate climate change.
- Understanding and detail of how the increased use of renewables might help mitigate against climate change, both in built structures and transportation.
- Explanation of, and the need for, improvements in urban design and how this may help mitigate against climate change.
- Explanation of how changing rural land use, grasslands, croplands and forested areas may help mitigate against climate change.
- Use of case studies.

Hints and tips

The question asks about mitigation and *not* about adaptation. When asked to 'assess the extent to which you agree', start with a clear statement about your thoughts on the matter. You then have to justify your assessment using facts and examples. Remember, alternative arguments are a useful way to illustrate the strength of your case. A summative statement at the end, bringing together your arguments, is important. There is a lot of information to cover so a clear structure is also important.

Question 12 example responses

Student A

Climate change mitigation refers to efforts to reduce or prevent greenhouse gas emissions. Since the increase of these gases is largely the responsibility of the human race, it is up to us to solve the problem. The various accords of the Paris Agreement 2016 put pressure on countries to design methods by which this can happen. All countries in the world now recognise the problem — the exception being the USA, which plans to revive its coal industry, a decision that can only add to the problem worldwide.

Carbon capture and sequestration is estimated to be able to capture up to 90% of all carbon dioxide emissions produced by the burning of fossil fuels, thus preventing them from entering the atmosphere. This is done by separating the gases and capturing the carbon dioxide, either before combustion, during combustion or after combustion. Once captured. the gas can then be transported by pipeline or tanker to the storage location. Storage takes place in old empty oil or gas fields. These are usually many kilometres below the surface. Geo-sequestration is where the gas is stored in suitable deep geological formations where the gas is liquified under pressure and injected into rocks that have an impermeable capping to prevent the gas from escaping. An example of this process in practice is the Boundary Dam plant in Saskatchewan, which hopes to capture a million tonnes of carbon dioxide a year. This gas will then be transported to a nearby oilfield, where it will be injected into oil-bearing strata and local salt formations.

Although there are many other methods of dealing with the problem, changes in the use of rural land is one that even relatively poorer countries can employ. Grass can store massive amounts of carbon by holding it in the soil. This can be utilised by not overstocking the land, adding manures and fertilisers, and improving existing grassland by irrigation to increase plant productivity. Rotation of crops improves the soil organic content (SOC), as does the use of improved crop varieties and a reduction in the amount of tillage. Forested areas and tree crops also store large amounts of carbon above and below the soil, so reforesting areas and planting trees on croplands will have a beneficial effect.

The aviation industry has realised that more fuel-efficient aircraft both help its bottom line and reduce the amount of greenhouse gases released into the atmosphere. The Airbus 380 is one such aircraft that has much improved fuel usage.

In urban environments, building design can help reduce carbon dioxide emissions through improved insulation, green roofs, use of grey water and renewable energy sources. Landfill sites could capture the gases and populations could be encouraged to recycle more. Recent innovations have been the introduction of a 5p charge for a plastic bag in shops and the proposal to charge an extra 25p for a plastic coffee cup.

Having created a situation where temperatures will rise between 2 and 4 degrees Celsius by 2100, the only way forward is for humans to put it right. In conclusion, therefore, agreement with the statement is necessary. The world's population simply cannot afford to wait and see what might happen or to hope for a natural solution, despite what Donald Trump thinks.

e The student gives a well-balanced argument using a clear structure. There is supporting evidence from a case study and a clear conclusion. The response is rather long, so the introduction could be shorter. Level 4: 17/18 marks

Student B

Mitigation of global warming refers to policies that are meant to slow, reduce or prevent climate changes. These policies aim to reduce carbon dioxide emissions, increase renewable energy production and usage, and increase carbon sinks by reforestation.

Emissions have to be reduced immediately if the world is to avoid irreversible climate change as temperatures continue to rise. It is estimated that at the current rate the Earth could see a rise of between 2 and 4 degrees Celsius by 2100. This could lead to extensive flooding, especially in low-lying countries like Bangladesh, and the flooding of some of the world's major cities, including London and New York.

Carbon sinks can be easily created by afforestation. This is something even poorer countries can do by encouraging farmers to plant trees on poorer-quality lands and in among their normal crops. Governments also need to prevent large-scale deforestation — something that the Brazilian government has failed to do.

Cutting emissions is difficult for some developing economies as economic growth leads to increases in emissions. This has been seen in China over the last 20 years, resulting in the country becoming the world's biggest polluter — a situation the Chinese government is already taking steps to correct. It now leads the world in the production of solar panels and renewable technologies.

In many western economies, emissions are actually falling as reduction becomes more politically acceptable. The exception is the USA, where the coal industry is being revitalised in order to 'make America great again'.

Humans have created the problem so humans must put it right, so I agree with the statement.

e **The student shows some knowledge of the issues but has neglected to use any detailed examples and fails to come to a satisfactory conclusion. In general, more depth and detail are required for Level 4. For both AO1 and AO2, the response achieves Level 3.** **Level 3**

Section B option: Coastal systems and landscapes

Set 1

Question 1 mark scheme

4 marks (AO1 = 4 marks). 1 mark for each valid point, with marks available for development.

Notes for answers

- A system is in dynamic equilibrium when its inputs and outputs of energy and matter balance. On coasts, systems remain in an almost steady state for long periods of time, with some short-term changes.
- Coastal systems adjust to these changes by a process of negative feedback.
- Beaches are a dynamic environment and constantly change in the short term but largely stay the same over long periods of time.
- Changes are caused by wave action, which alters depending on the weather and tidal conditions.
- A reduction in wave energy leads to the onshore movement of sediment. This results in a steeper beach profile, which causes greater backwash; when the swash equals the backwash, the beach profile is in equilibrium.
- Longshore drift might be mentioned.

Hints and tips

This question provides an opportunity to display your extensive use of technical language.

Question 1 example responses

Student A

Dynamic equilibrium refers to the changes in equilibrium conditions as a beach is being eroded and deposited. Changes in equilibrium conditions are affected by the wave action, the supply of sand and the direction the coastline is facing. Under stormy conditions, destructive waves will remove sand, making the beach flatter. Once the storm has passed the calmer conditions will produce constructive waves, which will add sand to the beach. Longshore drift will add material at one end of a beach while removing sand from the other end. All of this maintains a dynamic equilibrium.

e **The student identifies the main issues and gives some detail. There is good use of technical language and a clear understanding of process.** **4 marks**

Student B

Dynamic equilibrium is where a balance is maintained over time, with storms removing sand and calm conditions putting it back. A lot depends on the tides and on whether the beach faces into the wind or is sheltered.

e **The student has not got to grips with this question and has left out crucial details.** **1 mark**

Question 2 mark scheme

6 marks (AO1 = 2 marks, AO2 = 4 marks). Use the levels of response mark scheme as follows.

Level 2 (4–6 marks)

- **AO1** — Demonstrates clear knowledge and understanding of concepts, processes, interactions and change.
- **AO2** — Applies knowledge and understanding to the novel situation, offering clear analysis and evaluation drawn appropriately from the context provided. Connections and relationships between different aspects of study are evident, with clear relevance.

Level 1 (1–3 marks)

- **AO1** — Demonstrates basic knowledge and understanding of concepts, processes, interactions and change.
- **AO2** — Applies limited knowledge and understanding to the novel situation, offering basic analysis and evaluation drawn from the context provided. Connections and relationships between different aspects of study are basic, with limited relevance.

Notes for answers

- Biological weathering with roots breaking up the upper layers of rock.
- Upper layers of rock 'look' more prone to weathering; large blocks have fallen down on the right of the photograph.
- Lower layers have joints clearly shown, allowing rainwater in and causing mechanical freeze–thaw processes in winter.

- Effect of salt spray on minerals in the rock causing physical weathering. This includes:
  - salt crystallisation. This is when salt crystals are deposited in cracks and over time the salt accumulates and applies pressure to the crack.
  - wetting and drying. Clay-rich rocks are prone to expand when they are wet and contract when they dry. This results in cracks that are vulnerable to both freeze–thaw and salt crystallisation.
- Use examples and technical language.

Hints and tips

Divide the photograph up into four areas (top left, top right, bottom left, bottom right) for reference purposes — it will make it easier for you and your answer will be clearer for the examiner. This is also a good opportunity to use your fieldwork, if relevant.

Question 2 example responses

Student A

The cliff is made up of soft sedimentary rocks with the layers clearly shown. At the top of the cliff there are trees and shrubs, which will have encouraged biological weathering. The roots will have forced their way into the rock, forcing it apart and allowing water to get into the soil. This water may have been frozen, causing mechanical weathering or freeze–thaw, and the water may have reacted with minerals in the rock, causing chemical weathering.

All of this has caused mass movement and material has slid down the cliff face on to the beach. Lower down the cliff, where the rocks appear to be more solid but still in layers, there are obvious-looking joints in the rock, which again will allow water in and freeze–thaw weathering may be active during the winter months. On the right-hand side, some quite large blocks have fallen down. As this is at the coast, it is likely that minerals in the rocks may be affected by the salt spray during stormy weather. The reaction causes chemical weathering and the rock to weaken and be more likely to be eroded.

My coastal fieldwork was to Hunstanton where we saw evidence of these types of weathering on the cliffs, as well as mass movement where large amounts of materials had slid down, protecting the foot of the cliffs from further erosion.

e **The student uses good technical language and clearly understands the different processes at work. The reference to fieldwork could be improved by including some more detail.**

High Level 2

Student B

Weathering is where the rock is broken down in situ in a variety of ways. Here at the top of the cliff there is biological weathering, while lower down there is mechanical weathering. There may be some chemical weathering, but it is not very clear from this picture. The process of erosion is also possible and the cliff may be undercut in places, but it is not clear where the sea is from the picture. My coastal fieldwork looked at cliff erosion and we saw some examples of this and of weathering of various sorts.

e **The student starts well with some technical terms but drifts into erosion. Overall the answer lacks detail and does not make the best use of the resource.** **Level 1**

Question 3 mark scheme

6 marks (AO3 = 6 marks)

Completing the table and the calculations (3 marks)

Height of storm beach (m)	Shingle diameter (cm)	R_1	R_2	d	d^2
5.5	8.4	1	8	−7	49
8	7.7	2	7	−5	25
11	6.1	5	3	2	4
10	6.2	3.5	4	**−0.5**	0.25
10	7.5	3.5	6	−2.5	6.25
14	6.5	6	5	1	1
18	4.8	8	1	7	49
15	5.8	7	2	5	25
					$\Sigma =$ **159.5**

$$R_s = 1 - \left(\frac{6\sum d^2}{n^3 - n}\right) \qquad R_s = 1 - \left(\frac{957}{504}\right) \qquad R_s = \mathbf{-0.90}$$

Interpretation (3 marks)

- The R_S result of −0.9 exceeds the 0.05 and 0.01 significance levels (1 mark).
- This means that the null hypothesis can be rejected and that there is a significant relationship between beach height and sediment size (1 mark).
- It is a strong negative relationship — the lower the storm beach, the larger the sediment (1 mark).

Note: Allow 3 marks for a correct interpretation against an incorrect calculation of R_S.

Hints and tips

This question tests your ability to do simple calculations and comment geographically on the result.

Question 3 example responses

Student A

10	6.2	3.5	4	**–0.5**	0.25

$\sum d^2 = 159.5$

$R_s = 1 - \left(\frac{6\sum d^2}{n^3 - n}\right)$ $R_s = 1 - \left(\frac{957}{504}\right)$ $R_s = -\mathbf{0.90}$

The value for R_s is –0.90, which is greater than the critical value at the 0.05 and the 0.01 significance levels. This means the null hypothesis can be rejected and we can say with a 99% certainty that there is a strong relationship between shingle size and the height of storm beaches.

e The student provides three correct calculations and fully interprets the result with accuracy. **3 + 3 = 6 marks**

Student B

10	6.2	3.5	4	**–0.5**	0.25

$\sum d^2 = 159.5$

$R_s = 1 - \left(\frac{6\sum d^2}{n^3 - n}\right)$ $R_s = 1 - \left(\frac{957}{504}\right)$ $R_s = \mathbf{-0.9}$

The null hypothesis can be rejected.

e The student produces three accurate calculations and partially interprets the result but this interpretation is lacking in detail. **3 + 1 = 4 marks**

Question 4 mark scheme

20 marks (AO1 = 10 marks, AO2 = 10 marks). Use the levels of response mark scheme as follows.

Level 4 (16–20 marks)

AO1

- Detailed, relevant and appropriate knowledge and understanding of place(s) and environments.
- Full and accurate knowledge and understanding of key concepts.
- Detailed awareness of scale and temporal change where appropriate.

AO2

- Detailed, evaluative conclusion.
- Detailed application of knowledge and understanding throughout.
- Full evidence of links between knowledge and understanding.

Level 3 (11–15 marks)

A01

- Generally clear and relevant knowledge and understanding of place(s) and environments.
- Generally clear and accurate knowledge and understanding of key concepts.
- Generally clear awareness of scale and temporal change where appropriate.

A02

- Clear, evaluative conclusion.
- Generally clear application of knowledge and understanding.
- Generally clear evidence of links between knowledge and understanding.

Level 2 (6–10 marks)

A01

- Some relevant knowledge and understanding of place(s) and environments.
- Some knowledge and understanding of key concepts.
- Some awareness of scale and temporal change where appropriate. There may be a few inaccuracies.

A02

- Some sense of a conclusion.
- Some partially relevant application of knowledge and understanding.
- Some evidence of links between knowledge and understanding.

Level 1 (1–5 marks)

A01

- Very limited knowledge and understanding of place(s) and environments.
- Very limited knowledge and understanding of key concepts.
- Very limited awareness of scale and temporal change where appropriate. There may be a number of inaccuracies.

A02

- Very limited and/or unsupported conclusion.
- Very limited application of knowledge and understanding.
- Very limited evidence of links between knowledge and understanding.

Notes for answers

A01 Knowledge and understanding of the need for, and methods of, coastal management. For example:

- Shoreline management plans (SMPs) were introduced by the UK government in 1995. The coastline of England and Wales was divided into 11 sediment cells, each of which has an SMP designed to identify the most sustainable approach to managing the flood and coastal erosion risks to the coastline.
- Coastal management has four options:
  - Option 1: Hold the line — existing coastal defences are maintained but no new defences are set up.

- Option 2: Advance the line — new defences are built further out in the sea to reduce the stress on current defences and possibly slightly extend the coastline.
- Option 3: Retreat the line or managed realignment — move people out of danger zones and allow natural processes to take over.
- Option 4: Do nothing — the most sustainable option of just dealing with the effects of flooding and erosion as they come, or ignoring them.

- Detailed, place-specific examples should be given.

AO2 Application of knowledge and understanding of coastal management strategies to evaluate how sustainable they are. For example:

- Make clear links between coastal management approaches and their sustainability.
- Question the application of sustainability.
- Evaluate management schemes at various locations, such as the managed realignment at Abbotts Hall Farm.
- Show an awareness of scale and temporal changes.
- Give evidence of wider reading and/or reference to fieldwork.
- Provide a conclusion referring back to the question, which may well mention that sustainable is not totally possible, so we do the best we can.

Hints and tips

The quality of your answer will be determined by the way in which your points are supported by reference to a place or places that you have studied. Depth, detail and up-to-date material will also benefit your answer. You can use examples from outside the UK.

Using paragraphs will help isolate your ideas and give your answer a much better structure. Include a brief introduction, setting the scene for what follows, and an equally brief conclusion referring back to the question to round things off at the end.

Question 4 example responses

Student A

I agree with the statement. Some hard engineering schemes only have a limited lifespan and are very expensive, whereas some soft management schemes are not only cheaper and more affordable, but are able to adapt to changing circumstances as climate changes.

With rising sea levels and increased coastal urbanisation, the need for coastal management has become pressing. In 1995 the UK government established the idea of sediment cells — there are 11 covering the entire coast of England and Wales. Each one has a shoreline management plan and within them a variety of management approaches has been adopted.

In some areas where there are towns or important infrastructures, like the gas terminal at Easington, hard engineering has been used to hold the line. Sea walls, groynes, gabions and revetments are all hard structures designed to absorb and reflect the energy of the waves before they can attack the shore. They are not sustainable, however, as they require constant upgrading and repair.

'Retreat the line' means moving sea defences back and surrendering some areas of the coast to protect other areas. This has been undertaken at Rye Harbour in East Sussex, a scheme we saw on our fieldwork day at the coast there.

'Advance the line' means moving the defences forward to take pressure off existing defences. This type of defence can be seen at Sea Palling in Norfolk, where several offshore reefs have been constructed. A similar project is underway at Monster in the Netherlands, where sea defences are very important as much of the country is at or below sea level.

'Do nothing' is where defences are not maintained and eventually fall into disrepair, meaning the sea can break through. In some places this has been allowed to happen and, despite local controversy, areas of low-lying land that had been reclaimed from the sea have been allowed to flood at high tides. This will eventually create a salt marsh, which is a natural and sustainable form of defence.

These three approaches have a degree of sustainability, although the retreat and advance options imply that defence work will be undertaken and maintained into the future.

At Abbotts Hall Farm in Essex, managed realignment has taken the pressure off the coastal strip, which was suffering from coastal squeeze, and created a habitat for sea birds and salt-loving plants. At Medmerry in West Sussex, a similar scheme at a cost of £28 million was identified as being suitable for this type of defence. The cost was less than having to continually repair old defences, and it offers greater protection to 350 houses as well as to the area in general. This site also contains an area that is part of the UK Biodiversity Action Plan, which will provide a habitat for wildlife on an international level. In Porlock Bay, following a breach in the shingle ridge that protected the farmland behind, there has been no attempt to maintain the ridge as it was. Instead, the area behind the ridge floods regularly, creating a salt marsh that acts as a buffer zone. It was thought that maintaining the ridge would be to write a 'blank cheque'. Both of these schemes are as close to sustainable as any scheme gets, with a minimum of expense on continual repairs.

In conclusion, as well as providing protection for humans, thought must also be given to the wider environment when undertaking coastal management. What is good for Place A may well cause untold problems a few miles along the coast at Place B. Schemes must be cost-effective in the longer term and not just be a short-term fix, as was sometimes the case in the past. Schemes may also have to be implemented in the face of local opposition to satisfy the common good. In the final analysis, sustainability is never completely achievable and so a variety of approaches will provide as sustainable a result as possible.

e The student gives detail and place specifics that relate directly to the question and they explain how different approaches have been used to manage coastlines sustainably. Good references are made to fieldwork examples and to research done elsewhere. Level 4

Student B

Sustainability means providing for people now without affecting people in the future. Some approaches to coastal management are not sustainable, such as hard engineering solutions, but others, like soft engineering, are. Soft engineering includes such things as beach nourishment, managed retreat or doing nothing. By dividing the coastline into sediment cells, each with its own shoreline management plan, the government can decide what approach to take where. Some places need hard engineering, e.g. where people live or if there is an important road running along the coast. For some places, the 'do nothing' approach is all right because there is nothing that needs protecting, just farmland or not even that. The most sustainable approach is the 'managed retreat' method where the sea defences are moved further inland, allowing the land in front of them to flood and create a salt marsh. This is very sustainable as it does not need millions of pounds of work to maintain it once it is done. It also allows for a salt marsh area to grow so provides a habitat for sea birds and salt-loving plants. In Essex, a farm that was suffering from coastal squeeze had its defences moved to allow the sea in in order to protect the farm. This was cheaper than to keep repairing the defences and so is regarded as sustainable. So it can be seen that a variety of methods have to be used, with cheap methods balancing the costly hard methods.

e The student attempts to answer the question and has some limited understanding of the issues, although there is a marked lack of detail. No named examples are used, there are areas of confusion and the response is rather disjointed. The use of paragraphs would have helped to clarify the student's thoughts and therefore helped with the structure of the answer.

Level 2

Set 2

Question 5 mark scheme

4 marks (AO1 = 4 marks). 1 mark for each valid point, with marks available for development.

Notes for answers

Low-energy coasts:

- Waves are not powerful (for a significant part of the year).
- The rate of deposition exceeds the rate of erosion.
- Characteristic landforms include beaches and spits.
- Example: Dawlish Warren in Devon.

High-energy coasts:

- Waves are powerful (for a significant part of the year).
- The rate of erosion exceeds the rate of deposition.
- Characteristic landforms include headlands, cliffs and wave-cut platforms.
- Example: Seven Sisters coast in Sussex.

Hints and tips

This question provides you with an opportunity to display an extensive use of technical language.

Question 5 example responses

Student A

High-energy coasts are typically coasts that experience powerful waves for a large part of the year. As a result, the rate of erosion exceeds the rate of deposition. This gives rise to landforms such as are found along the Sussex coast at the Seven Sisters headlands: cliffs and wave-cut platforms.

With low-energy coasts, on the other hand, the waves are not as powerful and so the rate of erosion is less and the rate of deposition is much greater. This means that the coastline is typically one of beaches and spits. An excellent example of this is the coast at Dawlish Warren in Devon.

e The student identifies the main issues and has given some detail. There is good use of technical language and a clear understanding of processes. **4 marks**

Student B

High-energy coasts have rough seas and so get a lot of erosion, leaving cliffs and stacks behind due to the erosion. Old Harry Rocks are a good example.

Low-energy coasts are the opposite, with calm seas and no erosion so they have no cliffs but beaches from the deposition.

e The student has some basic ideas about the different types of coast but has left out crucial details and made some sweeping generalisations. **2 marks**

Question 6 mark scheme

6 marks (AO1 = 2 marks, AO2 = 4 marks). Use the levels of response mark scheme as follows.

Level 2 (4–6 marks)

- **AO1** — Demonstrates clear knowledge and understanding of concepts, processes, interactions and change.
- **AO2** — Applies knowledge and understanding to the novel situation, offering clear analysis and evaluation drawn appropriately from the context provided. Connections and relationships between different aspects of study are evident, with clear relevance.

Level 1 (1–3 marks)

- **AO1** — Demonstrates basic knowledge and understanding of concepts, processes, interactions and change.
- **AO2** — Applies limited knowledge and understanding to the novel situation, offering basic analysis and evaluation drawn from the context provided. Connections and relationships between different aspects of study are basic, with limited relevance.

Notes for answers

- The rock strata are dipping landwards, meaning that they are resistant to erosion and, in the photo, form a headland with bays on both sides.
- Wave refraction occurs — the wave fronts become distorted as they approach the coast due to sea bed friction.
- The orthogonals (lines of wave energy) become focused on headlands and diverge into the bay.
- Headlands are eroded from both sides.
- Erosion in the form of hydraulic action and abrasion picks out weaknesses in the rock to start the sequence of outcomes (wave-cut notch, cave, arch). Further erosion of the base of the arch, as well as sub-aerial weathering of the top of the arch, eventually causes collapse and leaves a stump. This erodes to a wave-cut platform.
- Include further examples, diagrams and references to fieldwork.
- Use technical language precisely.

Hints and tips

Divide the photograph up into four squares (top left, top right, bottom left, bottom right) for reference purposes — it will make it easier for you and your answer will be clearer for the examiner. This is also a good opportunity to use your fieldwork, if relevant.

Question 6 example responses

Student A

This is an example of where a headland sticks out into the sea. The rock seems to be more resistant than in the bays next to it. The rock also is dipping in towards the land which means that it forms steep cliffs.

As the waves approach the coast, they become refracted because they are slowed off the headland by shallow water. The waves become focused on the headland and attack it on both sides at the same time. Weaknesses in the rock, such as joints, faults and bedding planes, are picked out by the waves. Abrasion occurs at the base of the cliffs where rocks are hurled at the coastline by the waves. Also, the sheer force of the wave and compressed air open up faults and joints until a cave is produced.

In the photo, there is an arch formed where caves on each side of the headland have eroded towards each other. The top of the arch will eventually collapse under its own weight when it is weakened by frost action and biological weathering. This will form a stack like the one to the right of the headland. This will be worn down to a stump, uncovered only at low tide.

e The student uses good technical language and clearly understands the processes at work. There is clear reference to the resource and clear explanation of the formation of named landforms. High Level 2

Student B

In the photograph there is a small stack which has resulted from the erosion of a headland which is just to the left of the photo. There is also an arch in the middle of the photo and then a gap where an arch would have been.

What happens is the waves attack the headland as it sticks out into the sea from both sides. So when a cave forms it goes right through, forming a small arch. When the arch falls down, a stack is left behind, which might be quite big or small depending how rough the seas are. Other features like stumps and a wave-cut platform will also be made, but they are small and can be covered when the sea is rough, making them dangerous to ships.

e The student's answer makes good use of the photograph, but it lacks key points such as wave energy. Although some knowledge is shown, there is a general lack of detail. Level 1

Question 7 mark scheme

6 marks (AO1 = 2 marks, AO2 = 4 marks). Use the levels of response mark scheme as follows.

Level 2 (4–6 marks)

- **AO1** — Clear knowledge and understanding are shown.
- **AO2** — Knowledge and understanding are applied to offer clear analysis and evaluation.

Level 1 (1–3 marks)

- **AO1** — Basic knowledge and understanding are shown.
- **AO2** — Limited knowledge and understanding are applied to offer basic analysis and evaluation.

Notes for answers

A01 Knowledge and understanding of a GIS and an SMP:

- A GIS enables you to layer various sets of data to get a composite picture of an issue. It allows visualisation and analysis of data to show previously hidden relationships.
- Whenever the data sets are updated, changes can be rapidly seen and analysed.
- SMPs were introduced by the UK government in 1995. The coastline of England and Wales was divided into 11 sediment cells and each one has an SMP designed to identify the most sustainable approach to managing the flood and coastal erosion risks to the coastline.

A02 Application of knowledge and understanding in using a GIS to inform an SMP. For example:

- The data sets that could be layered include land use maps, rates of erosion, flood risk, historical maps, old newspapers, surveys, primary fieldwork data, geology.
- With these data, linear and aerial measurements can be taken to see how much erosion or deposition has taken place. Calculations can be made as to the rates of erosion or the frequency of floods, for example, which will aid the decision-making process about what measures should be taken and where.

Hints and tips

This question focuses on your understanding of how a geographical information system operates and your ability to explain how it can be used in a local shoreline management plan.

Question 7 example responses

Student A

Shoreline management plans (SMPs) were introduced by the government and they identify the most sustainable approach to managing the flood and coastal erosion risks to the coastline in the short term (0–20 years), the medium term (20–50 years) and the long term (50–100 years). A geographical information system (GIS) represents the locality on a computer using composite layers of information rather than several different maps.

In order to put together an SMP, a wide range of information has to be collected. Some of this will be primary data, such as fieldwork surveys, and some will be secondary data. Having established a local base map to use in a GIS, different layers of information can then be added. Roads, buildings and utilities can all be indicated in different colours on the GIS to relate to the geology and the position of the coastline.

Historical maps can also be added and from these measurements can be made, both linear and aerial, to see how much erosion or deposition has taken place. Calculations can be made as to the rates of erosion, etc., and what measures must be taken and where.

Primary data can be collected on a GPS and this information can be plotted directly onto the GIS, thus saving many hours copying the data over. A GIS can interrogate far more data far more easily, making the production of an SMP quicker and more responsive to local needs.

e The student clearly understands what is meant by an SMP and a GIS and has demonstrated how the technology can benefit the production of a local SMP. High Level 2

Student B

GIS is the technology that can help make an SMP more useful. A GIS can put layers of information on to a map base in order to help with decision-making and can do it far more quickly than in the past. SMPs are there to make decisions about coastal management for a large part of the coast, not just locally, so it is important to have as much information as possible. A GIS makes this possible as it can have lots of different layers in different colours so that people can see where the problems are.

Information collected by fieldwork can be put directly on to the map using a GIS and then secondary data like old maps can go on as well. This can show where erosion is taking place and you can measure from one map to another to see how much has happened.

All sorts of data like roads and drains can be added, as well as putting on the geology. The geology is important as it shows where erosion will happen quickest. In the past you would have had to use a separate geology map. In this way, it can all be done very quickly and a decision can be made with all the information available and not just some of it, like in the past.

e The student has a grasp of what is meant by an SMP and a GIS, but the answer lacks coherence and detail. **High Level 1**

Question 8 mark scheme

20 marks (AO1 = 10 marks, AO2 = 10 marks). Use the levels of response mark scheme as follows.

Level 4 (16–20 marks)

AO1

- Detailed, relevant and appropriate knowledge and understanding of place(s) and environments.
- Full and accurate knowledge and understanding of key concepts.
- Detailed awareness of scale and temporal change where appropriate.

AO2

- Detailed, evaluative conclusion.
- Detailed application of knowledge and understanding throughout.
- Full evidence of links between knowledge and understanding.

Level 3 (11–15 marks)

AO1

- Generally clear and relevant knowledge and understanding of place(s) and environments.
- Generally clear and accurate knowledge and understanding of key concepts.
- Generally clear awareness of scale and temporal change where appropriate.

AO2

- Clear, evaluative conclusion.
- Generally clear application of knowledge and understanding.
- Generally clear evidence of links between knowledge and understanding.

Level 2 (6–10 marks)

AO1

- Some relevant knowledge and understanding of place(s) and environments.
- Some knowledge and understanding of key concepts.
- Some awareness of scale and temporal change where appropriate. There may be a few inaccuracies.

AO2

- Some sense of a conclusion.
- Some partially relevant application of knowledge and understanding.
- Some evidence of links between knowledge and understanding.

Level 1 (1–5 marks)

AO1

- Very limited knowledge and understanding of place(s) and environments.
- Very limited knowledge and understanding of key concepts.
- Very limited awareness of scale and temporal change where appropriate. There may be a number of inaccuracies.

AO2

- Very limited and/or unsupported conclusion.
- Very limited application of knowledge and understanding.
- Very limited evidence of links between knowledge and understanding.

Notes for answers

AO1 Knowledge and understanding of what contributes to the formation of differing coastal landscapes. For example:

- Coastal landscapes develop over time. Short-term features include beaches, long-term features include headlands and bays.
- The relief of the coast is important — high and steep as opposed to flat and low-lying.
- The geology and structure at the coast will affect the development of features — whether it is discordant or concordant, a bays-and-headlands coastline or a Dalmatian coastline.
- Is the sea-level change positive or negative?

AO2 Application of knowledge and understanding to evaluate the effects of sea-level change on the formation of differing coastal landscapes. For example:

- Submergence results from a relative rise in sea level and produces drowned valleys and lengthens the coastline.
- Emergence results from a relative fall in sea level and creates raised beaches, relict cliff lines and offshore features such as bars.
- A complex history can result in both submergence and emergence features in the same area, e.g. raised beaches and fjords in Scotland.
- Long-term changes, e.g. eustatic, isostatic and tectonic changes, affect coastal landscapes.
- Short-term changes, e.g. tides and the monthly tidal range, spring to neap, will change beaches.
- Storm surges will result in a short-term rise in sea levels, which may have a profound impact on coastal landscapes.
- The concept of dynamic equilibrium linking processes to the landscapes produced.

- The effect of global warming and the thermal expansion of oceans on coastal landscapes in the future.
- Application to locations studied.

Hints and tips

Here is an opportunity to show wider reading and research into this topic and to appreciate the longer geological timeframe and the shorter daily tidal regime. Your use of technical language is important, supplemented by references to fieldwork and examples. To get to the higher levels, you need a clear summative statement of assessment of the 'importance of the role' of sea-level change.

Question 8 example responses

Student A

Sea-level changes happen both in the short term and in the long term, but changes to coastal landscapes will depend on the nature of the sea-level change. A major factor will determine whether the change is a positive one, with the level rising, or a negative one, with the level falling. If sea level is rising, the lower reaches of river valleys may well be flooded, along with other low-lying areas. The drowned valleys or rias of southwest England and Ireland were formed in this way. Steep valley sides become cliff-like slopes and the coastline becomes a lot longer. Where the coastline was glaciated before the sea-level rise, the deep U-shaped valleys become the fjords of western Scotland, Norway and British Colombia. Falling sea levels create raised beaches and relict cliff lines, which can again be found in Scotland, on the east coast. This shows that sea level changes are very complex, meaning that examples of both falling and rising levels can be found in the same area. Our fieldwork on the Isle of Arran demonstrated this, with all these features being seen.

Another factor, independent of sea-level change, is the geology at the coastline. If the geology is concordant, a different set of features will be formed from those at coastlines where the geology is discordant. Discordant coasts present hard and soft rocks to the coast. This means that if the sea level is rising, the softer rocks will be eroded far more quickly, resulting in a bay-and-headland coast as the harder rocks prove more resistant to the erosion. In southern England, sea levels are rising and this type of coastline is common in places like Swanage Bay. Concordant coasts, where the geology and structure run parallel to the coast, will give rise to a Pacific or Dalmatian coastline, with lines of islands running along the coast and the intervening valleys drowned. Another good example is the coast off the British Colombian mainland.

The last ice age caused sea levels to fall at first as the ice covered the land, and then rise as the ice melted. This rise was offset by the land also rising as it readjusted to the removal of the weight of the ice. This makes the picture very complicated. Tectonic changes can also play a part. At Campo Flegrei in Italy, a recent study showed that there had been a rapid uplift of 11 cm between January 2012 and June 2013, and in December 2012 a 3 cm uplift had occurred.

Short-term changes will be related to the tidal activity on a daily and monthly basis. The tides will have an effect on small-scale features such as beaches as the sediment is moved within the sediment cell. Surges associated with storms raise the sea level locally so will also have an effect. Tropical storms can overrun dune systems along the east coast of the USA, as happened with Hurricane Hugo when it hit North Carolina in 1989.

From the examples offered, it is clear that the coastal landscape is a dynamic environment that reacts to the forces at work on it and responds in time to create the landforms we see. However, this equilibrium may well be upset as global warming and the thermal expansion of the oceans continue to pose a threat to low-lying areas of the world, such as Bangladesh, the east coast of the United States and the island nations of the Pacific. This will play a major role in the future development of the landscape.

e **The student provides detail and place specifics that relate directly to the question and explains how differing coasts have evolved. Good references are made to fieldwork examples and to research done elsewhere.** **Level 4**

Student B

Sea-level changes happen every day as the tides come and go, but the effect they have on the coastal landscape is quite minor, just rearranging the sand on the beach. Unless there is a storm surge, in which case the beach might get completely washed away along the coast, causing a different landscape in two places — where the sand was and where it is now. The main impact of sea level change is due to the ice age. When it was happening, sea levels were very much lower than they are now, and then, when the ice melted, the sea levels rose and drowned any features that might have been developed. But the ice had pushed the land down as well so, although there was a rise in sea level causing the formation of rias and fjords in parts of Britain, the land also came up, causing raised beaches to form. Rising sea levels are causing problems in low-lying places like Bangladesh and Kiribati due to climate change and the expansion of the oceans, and these islands are now flooding on a regular basis, changing the coastal landscape. Coastal defences have to be put in and this changes the landscape as it does not look as nice. If the coast is rocky and high, the rising sea levels do not have so great an effect other than flooding the valleys, but low ground will have extensive flooding and the coast will actually get longer. Then these valleys fill up with mud and the coast gets shorter again.

e **The student attempts to answer the question and has some limited understanding of the issues, although there is a marked lack of detail and the necessary technical language. No detailed examples are used and there are areas of confusion so the response is rather disjointed. The lack of paragraphs only makes the confusion greater and hides any structure that there might have been. There is no real discussion in the answer as it is largely descriptive.** **Level 2**

Section C option: Hazards

Set 1

Question 1 mark scheme

6 marks (AO1 = 2 marks, AO2 = 4 marks). Use the levels of response mark scheme as follows.

Level 2 (4–6 marks)

- **AO1** — Demonstrates clear knowledge and understanding of concepts, processes, interactions and change.
- **AO2** — Applies knowledge and understanding to the novel situation, offering clear analysis and evaluation drawn appropriately from the context provided. Connections and relationships between different aspects of study are evident, with clear relevance.

Level 1 (1–3 marks)

- **AO1** — Demonstrates basic knowledge and understanding of concepts, processes, interactions and change.
- **AO2** — Applies limited knowledge and understanding to the novel situation, offering basic analysis and evaluation drawn from the context provided. Connections and relationships between different aspects of study are basic, with limited relevance.

Notes for answers

Conditions that can lead to the formation of a tropical cyclone:

- An oceanic location with sea temperatures above 27°C — this provides a continuous source of heat in order to maintain rising air currents.
- An ocean depth of at least 70 m — this moisture provides latent heat; rising air causes the moisture to be released by condensation and the continuation of this drives the system.
- A location at least 5° north or south of the Equator in order that the Coriolis force can bring about the maximum rotation of the air (the Coriolis force is weak at the Equator, meaning that a circular air flow will not develop.
- Low-level convergence of air in the lower atmospheric circulation system — winds have to come together near the centre of the low-pressure zone.
- Rapid outflow of air in the upper atmospheric circulation — this pushes away the warm air that has risen close to the centre of the storm.

These conditions do not mean that a tropical cyclone will automatically develop, however.

Hints and tips

You must show that you have made use of Figure 9. To access higher marks in Level 2 you must make clear links between the ocean/climate conditions and the development of a tropical cyclone.

Question 1 example responses

Student A

Hurricanes and typhoons are the same thing but are just called different names in different parts of the world. The conditions under which they form are therefore the same. They form a little way away from the Equator, between 5° and 15° north or south. This is because there is no Coriolis force at the Equator and the Coriolis force is one of the requirements for tropical cyclone formation.

The conditions required were identified by Musk (1988), who came up with seven prerequisites. The first and probably the most important is a large body of warm water at 26°C and extending down for 70 m. There needs to be a strong Coriolis force to set up the winds in a circular pattern, building on a disturbed area in the lower atmosphere. These winds should then all be blowing in the same direction. To create the cyclone, the upper air must be flowing outwards and removing heat, while at ground level moist, unstable air must be drawn in.

An eye will form in the centre as a result of adiabatic warming, but the causes of this are not fully understood.

These conditions might, or equally might not, lead to the development of a tropical cyclone.

ⓔ **The student comprehensively covers the conditions required for the formation of a tropical cyclone, with good use of technical language.** **High Level 2**

Student B

A hurricane and a typhoon are really just the same thing and they are formed in the same way, over the sea near the Equator. The water in the sea needs to be quite hot, around 30°C, so that the air above it rises. The air then starts to spin because the Earth is rotating and it spirals up to the top of the atmosphere where it spreads out, forming the layer of clouds that you get with a hurricane. This air is unstable and humid so it produces lots of rain and high winds, which can cause severe damage, especially in less developed countries. There is an eye in the middle where the conditions are calm with no rain.

e **The student has some basic knowledge but has not engaged with the necessary language required, and towards the end has wandered off the point of the question. High Level 1**

Question 2 mark scheme

9 marks (AO1 = 4 marks, AO2 = 5 marks). Use the levels of response mark scheme as follows.

Level 3 (7–9 marks)

- **AO1** — The answer shows detailed knowledge and understanding.
- **AO2** — Knowledge and understanding are appropriately applied. Connections and relationships are made between the different aspects studied, which are fully developed and completely relevant. The evaluation is detailed and well supported. The answer provides a well-balanced and coherent argument.

Level 2 (4–6 marks)

- **AO1** — The answer shows clear knowledge and understanding, but there may be some minor inaccuracies.
- **AO2** — Clear knowledge and understanding are appropriately applied. Connections and relationships are made between the different aspects studied, which have some relevance. The evaluation is clear and is supported with appropriate evidence. The answer provides a clear but partial argument.

Level 1 (1–3 marks)

- **AO1** — The answer shows basic knowledge and understanding, with limited relevance and there are inaccuracies.
- **AO2** — Limited knowledge and understanding are applied. The connections and relationships made between the different aspects studied are basic, with limited relevance. The evaluation is basic. The answer provides a basic argument.

Notes for answers

AO1 Knowledge and understanding of the effects of a storm:

- A storm will have primary and secondary effects.
- The effects can be economic, environmental, social and political.
- The effects can be short, medium and long term.
- There will need to be a range of responses to all of these different effects.

AO2 Application of knowledge and understanding of the effects of a storm in the two different locations:

- Many effects of a storm will be the same in both countries, e.g. storm surges may affect low-lying coastal areas in both countries, and buildings and infrastructure can be destroyed in both countries.
- However, New Port Richey, Florida is located in the USA, a high-income country (HIC), whereas Temerloh is in Malaysia, which is a low- or middle-income country (LIC/MIC). Therefore, the abilities of the two countries to deal with the effects of a storm are different.
- Storms bring destructive high winds that, in LICs particularly, can destroy flimsy housing, making people homeless. In HICs, people can afford to build stronger houses to withstand the high winds and so are less likely to be made homeless.
- HICs will have services in place to prepare for and deal with disruption to, for example, infrastructure such as roads, bridges and utilities. This disruption will create a more significant problem in an LIC.
- Water pollution from sewage and the bodies of animals and people will be more of a problem in LICs due to lack of an alternative clean water supply. HICs will have rapid response units to bring in water from elsewhere (e.g. FEMA in the USA).
- Early-warning systems and evacuation plans will be in place in HICs to move people to places of safety. These may not be available in LICs.
- People in LICs often live in more dangerous places because the land is cheap — nobody else wants to live there. As such, they are more at risk from the effects of a storm, e.g. the risk of landslides in heavy rain.
- Storage facilities for crops in LICs may be destroyed by floods, leading to famine later in the year.
- In LICs, children who are orphaned may have to leave school to care for younger siblings; children losing their education will have a long-term effect on the economy.

Hints and tips

The key to this question is that the ability of each country to mitigate the effects of a storm is related to its stage of development.

Question 2 example responses

Student A

Both these areas are subject to annual storms, which can have a devastating effect on the people and landscape. Some of the impacts will be similar and some will be peculiar to that area, but the major difference is that New Port Richey is in the USA, an HIC, and Temerloh is in Malaysia, which is just becoming an MIC although many of its people are living in LIC conditions.

The immediate effects will be primary — the effect of the storm — but some effects will be secondary and will develop later as a by-product of the storm. Damage to property will be far greater in Malaysia as the houses of the poorer people will be flimsy and unlikely to be able to stand up to the destructive nature of winds blowing at 60 m/s. In Florida, buildings will be hurricane-proofed and built in such a way as to withstand these winds. A secondary effect in an LIC is that the price of building materials will go up as people try to rebuild their houses — they will be unable to afford these prices.

The economic effects will be felt far harder in the LIC, although the cost of the storm will be much higher in the USA. The people of Malaysia have so little and it could all be lost, while the residents of Florida may also lose everything but are likely to be insured, unlike their Malaysian counterparts. Environmentally, the effects will be greater in Malaysia as water supplies are likely to be polluted (by sewage, debris and the bodies of people and animals) and there is no agency to supply clean water, like FEMA in the USA. Deaths from waterborne diseases such as cholera will therefore be a secondary effect of the storm.

The socio-political effects are not so clear. President George Bush was criticised for not visiting the city of New Orleans immediately after Hurricane Katrina struck, so politicians need to be seen to be doing something. Society is affected in LICs, as parents may drown trying to save their children. The children are then left as orphans and have to leave school to look after younger siblings. The loss of education will be another secondary effect but one that will have a long-term effect on the country. Warning systems are in place in the USA to let people know that a storm is on its way, and there are clear evacuation routes and centres to go to. This is not the case in countries like Malaysia because they simply cannot afford them.

In the USA, millions of dollars-worth of damage will be caused, but the economy is so strong that the country soon recovers. Recovery in Malaysia, on the other hand, could take many years, meaning that the effects remain for much longer.

Therefore, although the effects of storms are broadly similar wherever they occur, the ability of a country to cope before and after the event will determine the actual effects that the storm has.

e **The student recognises the differences in impacts due to the difference in the countries' stages of development. There is evidence of wider research and some useful detail is included.** **High Level 3**

Student B

The two countries are at different places on the demographic transition model. The USA is at stage 4 but Malaysia is at stage 2. This means that Malaysia will be unable to cope as well as the USA. In the USA they will have an early-warning system so people can leave their homes and go somewhere safer. In Florida, there are special evacuation routes signposted so that there are no traffic jams on the day. Malaysia cannot afford to do this. So more people are at risk of drowning or being killed in Malaysia. Houses in the USA are built stronger so that they do not blow away in the storms, but huts made of straw and other natural materials in Malaysia will.

People will not be able to get about in Malaysia because the roads will be flooded and the bridges might be out, so, economically, this will be a problem for them. The USA has stronger bridges and other ways of transporting goods so they will be all right. Rescue efforts will be good in the USA and clean water in bottles can be supplied so that there is no risk of disease, but in Malaysia there will be a lot of disease, such as cholera, from polluted water.

In the short term, the USA will soon be up and running again because people have insurance and the government is very rich, but it could take years for Malaysia to recover because the people are poor and not insured and the government has no money to spend on things like this, so there will be long-term effects.

e **The student has some vague ideas but they are not well expressed or developed with any detail. There are a number of generalisations and a few inaccuracies.** **Middle Level 2**

Question 3 mark scheme

9 marks (AO1 = 4 marks, AO2 = 5 marks). Use the levels of response mark scheme as follows.

Level 3 (7–9 marks)

- **AO1** — The answer shows detailed knowledge and understanding.
- **AO2** — Knowledge and understanding are appropriately applied. Connections and relationships are made between the different aspects studied, which are fully developed and completely relevant. The evaluation is detailed and well supported. The answer provides a well-balanced and coherent argument.

Level 2 (4–6 marks)

- **AO1** — The answer shows clear knowledge and understanding, but there may be some minor inaccuracies.
- **AO2** — Clear knowledge and understanding are appropriately applied. Connections and relationships are made between the different aspects studied, which have some relevance. The evaluation is clear and is supported with appropriate evidence. The answer provides a clear but partial argument.

Level 1 (1–3 marks)

- **AO1** — The answer shows basic knowledge and understanding, with limited relevance, and there are inaccuracies.
- **AO2** — Limited knowledge and understanding are applied. The connections and relationships made between the different aspects studied are basic, with limited relevance. The evaluation is basic. The answer provides a basic argument.

Notes for answers

AO1 Volcanoes pose a range of hazards:

- Lava flows affect areas on the sides of mountains.
- Pyroclastic flows can affect large areas, often several kilometres from the mountain.
- Lahars can extend down valleys for many kilometres.
- The effects of toxic gases are felt close to the mountain before they disperse into the atmosphere.
- Explosive blast.

AO2 How effectively volcanic hazards can be managed depends on:

- The variation in the scale or nature of the hazards, e.g. lava flows are easier to manage than lahars.
- The quality of the observing and monitoring systems used.
- Whether hazard mapping and geographical information systems are used — these can help predict hazards but are never 100% accurate.
- Whether evacuation plans to places of safety are in place — although secondary problems of disease can occur in camps.
- Whether short-term management plans are in place — warning systems, rescue and evacuation services.
- Whether long-term management plans are in place — monitoring and preparedness planning.
- The ability of a country to put into place long-term and short-term planning measures is affected by its level of development.

Hints and tips

The question asks you to 'assess the extent to which…', so it is useful to acknowledge that some hazards are easier to manage than others and that the effectiveness of individual management strategies will depend on a country's level of development. References to examples are needed to reach at least a Level 2 answer.

Question 3 example responses

Student A

The eruption on Mount Etna is a good example of volcanic hazard management. Volcanoes present a range of hazards including lava flows, pyroclastic flows, lahars, ash deposits, and toxic gases, all of which need managing.

Managing lava flows depends on the sort of lava — it might only flow a short distance down the hillside. In the past, the lava flows on Mount Etna have been managed by putting concrete blocks in the path of the lava to divert it away from houses and by dropping cold water on it from army helicopters to cool the lava down and therefore slow the flow. A similar technique of spraying cold water on lava was also used in Iceland at Heimaey to stop the lava destroying the town and harbour. In the 2002 eruption at Mount Nyiragongo in the Democratic Republic of the Congo, the lava flowed up to speeds of 90 km/hr, which made management techniques very difficult.

Pyroclastic flows can move at up to 300 km/hr and travel some distance from the volcano. Managing this hazard is less easy, but the preparation of hazard maps and planning zones can help remove people from the high-risk zones by stopping houses being built in these areas.

Managing lahars is even more difficult, if not impossible. These are flows of hot mud, ash and water that travel down river valleys as far as is possible. Again, hazard mapping using GIS applications might be a way forward, linked to planning zones. Maps do not always work, however, as was seen at Mount St Helens in the USA, which erupted sideways rather than straight up as expected.

Toxic gases soon get dispersed, but close to the mountain they do cause deaths. Large eruptions like Pinatubo in the Philippines in 1991 put huge amounts of gases into the upper atmosphere, creating sulfuric acid that had serious effects around the world for several years. This type of hazard cannot be effectively managed.

The effective management of volcanic events will largely depend on the resources available to the country in question. An LIC such as the Democratic Republic of the Congo will have few resources to use in order to plan and protect its people. It might be able to move people away from the danger zone, but the camps they would have to stay in would have a high risk of disease. HICs like Italy do have the resources — GIS to make hazard maps, evacuation routes and procedures, emergency search and rescue teams, stores of clean water and temporary housing all mean that Italian volcanic hazards can be effectively managed.

e The student shows substantial knowledge and understanding of the various factors involved in managing volcanic hazards, and gives plenty of examples. High Level 3

Student B

The management of volcanic hazards is difficult as the hazards are so varied and what you plan for might not happen. This was the case at Mount St Helens, which everyone thought would erupt straight up but instead went sideways, making all the plans useless.

Hazards like lava flows are easier to manage. For Etna, they put concrete blocks in the path of the lava to divert it away from the farms and houses. In Iceland, they pumped cold water from the sea onto the lava to cool it down and stop it flowing into the harbour and destroying the fishing boats. This was effective management and a method that can be used again. Other hazards like pyroclastic flows and lahars are more difficult to manage as they move so fast and travel over longer distances. The only way is to make sure that no one lives in the way of them by having hazard maps, early-warning systems and evacuation routes so people know where to go in the event of a hazard.

Countries that are poor may not have the means to make these maps and the people might not want to leave their farms to go to refugee camps. They know that the camps are likely to be poorly equipped and lack good facilities and that they might die of disease. Rich countries do not have this problem as they have temporary houses that can be put up, and they have stocks of water and emergency rescue teams who can rescue people.

e The student shows some knowledge and understanding of the management of hazards, and there is an attempt to use examples. Middle Level 2

Question 4 mark scheme

20 marks (AO1 = 10 marks, AO2 = 10 marks). Use the levels of response mark scheme as follows.

Level 4 (16–20 marks)

AO1

- Detailed, relevant and appropriate knowledge and understanding of place(s) and environments.
- Full and accurate knowledge and understanding of key concepts.
- Detailed awareness of scale and temporal change where appropriate.

AO2

- Detailed, evaluative conclusion.
- Detailed application of knowledge and understanding throughout.
- Full evidence of links between knowledge and understanding.

Level 3 (11–15 marks)

AO1

- Generally clear and relevant knowledge and understanding of place(s) and environments.
- Generally clear and accurate knowledge and understanding of key concepts.
- Generally clear awareness of scale and temporal change where appropriate.

AO2

- Clear, evaluative conclusion.
- Generally clear application of knowledge and understanding.
- Generally clear evidence of links between knowledge and understanding.

Level 2 (6–10 marks)

AO1

- Some relevant knowledge and understanding of place(s) and environments.
- Some knowledge and understanding of key concepts.
- Some awareness of scale and temporal change where appropriate. There may be a few inaccuracies.

AO2

- Some sense of a conclusion.
- Some partially relevant application of knowledge and understanding.
- Some evidence of links between knowledge and understanding.

Level 1 (1–5 marks)

AO1

- Very limited knowledge and understanding of place(s) and environments.
- Very limited knowledge and understanding of key concepts.
- Very limited awareness of scale and temporal change where appropriate. There may be a number of inaccuracies.

AO2

- Very limited and/or unsupported conclusion.
- Very limited application of knowledge and understanding.
- Very limited evidence of links between knowledge and understanding.

Notes for answers

AO1 Knowledge and understanding of the potential for hazards, for example in the Philippines:

- The Philippines lies on a major plate boundary, that of the Pacific and Eurasian plates. This is a destructive margin, with the Eurasian plate being subducted. Volcanoes and earthquakes are associated with the zone of subduction here, known as the 'Ring of Fire'.
- The northern and eastern coasts of the Philippine islands face the Pacific, which is the world's most tsunami-prone ocean.
- The Philippines lies in a major typhoon belt — around 19 tropical storms enter the region in a typical year and of these usually six to nine hit the islands. High winds lead to destruction of crops and infrastructure. Flooding is common.
- The landscape of the Philippines has steep, mountainous slopes that are prone to landslides, especially during tropical storms.
- In contrast, countries like the UK are away from tectonic boundaries, rarely suffering serious earthquakes and never experiencing volcanoes. The temperate climate has few extremes, but there is an increasing hazard of wildfire with drier summers.

AO2 Application of knowledge and understanding to show how the chosen hazardous country has become one of the most hazard-prone in the world, for example the Philippines:

- The Philippines has a long history of seismic and volcanic events. It has 53 active volcanoes and the eruption of Mount Pinatubo in 1991 was the second largest eruption recorded in the twentieth century. Major earthquakes occurred in 1990, 1994 and 2013.
- The magnitude of earthquakes in the Philippines is generally on the high end of both the Richter scale and the modified Mercalli scale, meaning they are strong enough to cause major damage and destruction. Poorly built houses will not be able to withstand these.

- The Philippines is the most exposed country in the world to tropical storms. Monsoons and typhoons (called *bagyo*) affect thousands of people, causing flooding and landslides.
- El Niño and La Niña also influence conditions in the Philippines. La Niña is associated with anomalies in rainfall, temperature and tropical cyclone activities. El Niño conditions, on the other hand, favour the formation of tropical cyclones over the western Pacific, increasing the number of these events..
- Human activities such as deforestation contribute to the occurrence of landslides. People live on the steep, mountainous slopes to farm because there is such pressure on the land from a growing population.
- Poverty and a growing population also mean that more people are living in low-lying coastal areas, where poorly constructed housing and inadequate evacuation plans leave them at risk from tropical storms.
- In contrast, the UK has a temperate climate with few extremes. Wildfires are on the increase with changing climate, but are still limited in extent. The UK lies far from the nearest plate boundary and rarely experiences a serious earthquake.

Hints and tips

Here is an opportunity to show wider reading and research into this topic. Your use of technical language is important, supplemented by references to specific events.

Question 4 example responses

Student A

Although all countries face a mix of hazards, some countries face more than their fair share. The Philippines is one of these and can be classified as a hazard hotspot. The reasons are obvious. The country is made up of over 7600 islands so faces all sorts of risks from the sea, but the major risk is tectonic. The islands lie at a major plate boundary between the Pacific and Eurasian plates. This is a destructive margin with the Eurasian plate being subducted, creating volcanoes such as Pinatubo on the Pacific plate. The plates do not slide smoothly past each other; instead they stick so that pressure builds up and then there is sudden movement, creating an earthquake. Earthquakes often occur as a signal to volcanic activity. In 1990 the Philippines had an earthquake measuring 7.8 on the Richter scale; this resulted in 1666 deaths and US$10 billion of damage. Other recent earthquakes include one measuring 7.1 in 1994 and one measuring 6.7 in 2012. There are very few areas in the Philippines where the modified Mercalli scale is below a value of 8. A Mercalli scale 8 means very strong and considerable damage will be done to poorly built houses, such as those in shanty towns. There is a strip that runs through the country along the line of the plate boundary where the scale is between 9 and 10 — a scale 10 is where even well-built buildings will collapse.

There are 53 active volcanoes in the country, the best known being Pinatubo, which erupted catastrophically in 1991. It lost 300 m from its summit in this eruption and sent material into the upper atmosphere that circulated for two years. Only 300 people died, but large areas of farmland were destroyed by being covered by lahars and ash. This was a disaster for the farmers and their families. Overall, the economic damage was over US$700 million.

These events and the heavy rainfall that the islands suffer lead to landslides on the steep mountain slopes. People live on these slopes, both to farm and because there is such pressure on the land from a growing population. Landslides occur every year, some with a huge loss of life because the farmers remove the natural forest and plant cash crops like coconuts, which are a very shallow-rooted tree and so do not bind the soil together as well.

The monsoons and typhoons bring huge amounts of rain and high winds. These damage buildings and displace large numbers of people. In August 2016 over a quarter of a million people were forced to leave their homes and the city of Dagupan was declared a state of calamity.

The influences of El Niño and La Niña lead to further problems of heavy rain and/or drought, which add to the disaster list. The heavy rainfall can interact with the volcanic and seismic hazards to produce landslides and lahars.

It is clear that the Philippines fully justifies its place in the list of most hazard-prone nations of the world.

In contrast, somewhere like the UK is far from any major plate boundary and so has no volcanoes and experiences very few serious earthquakes. The UK has a temperate climate with few extremes and, although there is flooding, it is much less extensive than in the Philippines.

e The student shows sound knowledge of the hazards that affect the Philippines, and their causes and effects. Reference is made to particular events, with supporting detail and use of technical language. Level 4

Student B

My example of a multi-hazardous country is the Philippines. It is one of the six most hazard-prone countries in the world. It is right above a plate boundary between the Pacific and Eurasian plates, where subduction is taking place between the plates. This leads to the formation of volcanoes and the Philippines has 53 active ones. The best known is Pinatubo, which blew up in 1990, sending masses of material into the atmosphere, but also caused lahars and ash flows that destroyed people's homes and farms. This was an economic disaster as well as a volcanic one. With volcanoes come earthquakes, often just before a volcano erupts, and the Philippines has many active volcanoes so it has a lot of earthquakes as well. The last big one was in 2013 — a 7.2 quake that killed 222 people. There were other big ones in 1994 (7.1) and in 1990 (7.8), which was one of the most powerful and killed 1666 people. This last one came at the same time as the Pinatubo eruption and they might have been linked.

The steep mountain sides are affected by earthquakes and so landslides are common, which kill large numbers of people. The farmers grow coconuts, which are shallow-rooted plants that do not bind the soil together as well as the tropical rainforest that they have cleared. So by growing these, farmers make things worse when there is a landslide.

Typhoons and very strong monsoon winds and rain are another hazard for the Philippines, causing more landslides and so more people have to be looked after in temporary camps. The monsoon comes every year and the typhoons are connected with El Niño or La Niña and so cannot be avoided.

The country also gets tsunamis, which are caused by earthquakes in the Philippines or in other countries around the Pacific, like the recent one in Chile. With all this going on, the Philippines is a real hotspot for hazards.

In contrast, the UK has fewer hazards. It has no seismic or volcanic hazards and the climate is not extreme, though there are sometimes floods and wildfires.

e The student has some knowledge of hazards in this region, but does not provide the necessary accuracy or detail all the time. Some generalisations are made and the language falls below the standard required at this level. Middle Level 2

Set 2

Question 5 mark scheme

6 marks (AO1 = 2 marks, AO2 = 4 marks). Use the levels of response mark scheme as follows.

Level 2 (4–6 marks)

- **AO1** — Clear knowledge and understanding are shown.
- **AO2** — Knowledge and understanding are applied to offer clear analysis and evaluation.

Level 1 (1–3 marks)

- **AO1** — Basic knowledge and understanding are shown.
- **AO2** — Limited knowledge and understanding are applied to offer basic analysis and evaluation.

Notes for answers

- Figure 12 shows a destructive plate margin with two converging plates.
- The oceanic plate is subducting beneath the continental plate.
- Friction creates an oceanic trench.
- The earthquake sequence goes from shallow through intermediate to deep.
- The leading edge of the oceanic plate melts at the Benioff zone.
- Magma rises through the crust to the surface as plutons. The plutons create volcanic landforms, both intrusive and extrusive.
- Offshore eruptions create island arcs such as the West Indies, Guam and the Marianas.

Question 5 example responses

Student A

The figure represents two converging plates at a destructive plate margin. The oceanic plate is being subducted under the continental plate, like the situation on the west coast of South America. The friction between the two plates creates a deep-water trench in line with the edge of the descending plate. Earthquakes may also occur but only shallow quakes and offshore. Shallow earthquakes are often the most destructive, as seen in Italy in 2016. As the plate continues to descend, it will create further intermediate earthquakes up to 50 miles down, but these do less damage. The deep earthquakes will occur in the Benioff zone up to 400 miles down, in the area where the leading edge of the plate is now beginning to melt. The melting material then rises as magma plutons through the crust to form intrusive features like batholiths and extrusive features like volcanoes. In South America, mountains like Cotopaxi and Nevado del Ruiz are good examples of this. If the molten material comes up in the sea, you get island arcs like the West Indies.

The student demonstrates a clear understanding of the processes at work and includes some examples to add detail. There is excellent use of technical language. High Level 2

Student B

The diagram shows a destructive plate boundary where one plate goes under the other. As it goes down there are various earthquakes, some near the surface, some halfway down and some at the bottom where the plate is melting. These are deep earthquakes that are highly destructive when they occur. The melting of the plate then sends molten material up to the surface, where it forms volcanoes or sometimes features below ground if it cannot break through. Different sorts of volcano can form but they are usually explosive types like Pinatubo.

e The student produces an answer that is lacking in detail and accuracy. It is a superficial account showing an almost complete lack of technical language. Middle/high Level 1

Question 6 mark scheme

9 marks (AO2 = 5 marks, AO3 = 4). Use the levels of response mark scheme as follows.

Level 3 (7–9 marks)

- **AO2** — Knowledge and understanding are appropriately applied. Connections and relationships are made between the different aspects studied, which are fully developed and completely relevant. The evaluation is detailed and well supported. The answer provides a well-balanced and coherent argument.
- **AO3** — Detailed analysis of the quantitative evidence. Detailed connections made.

Level 2 (4–6 marks)

- **AO2** — Clear knowledge and understanding are appropriately applied. Connections and relationships are made between the different aspects studied, which have some relevance. The evaluation is clear and is supported with appropriate evidence. The answer provides a clear but partial argument.
- **AO3** — Clear analysis of the quantitative evidence provided, which makes appropriate use of evidence in support. Clear connection(s) between different aspects of the evidence.

Level 1 (1–3 marks)

- **AO2** — Limited knowledge and understanding are applied. The connections and relationships made between the different aspects studied are basic, with limited relevance. The evaluation is basic. The answer provides a basic argument.
- **AO3** — Basic analysis of the quantitative evidence provided, which makes limited use of evidence in support. Basic connection(s) between different aspects of the evidence.

Notes for answers

AO3

- There are far more human-caused fires than lightning-caused fires.
- The total number of fires fluctuates at around 8000. The highest number was just over 10,000 in 2008 and the lowest number was 6500 in 2010.
- Although lightning-caused fires have decreased in number since their peak in 2001 (when there were approximately 3000), there is no real trend in the total number of fires for either lightning-caused or human-caused fires.

- The area burnt averages about 2400 km^2. The number peaked at just over 5800 km^2 in 2008, the year with the largest number of fires. The lowest number was 500 km^2 burnt in 2010.
- The number of fires does not necessarily correlate to the area burnt. For example, in 2004 and 2009 both had a high number of fires but a relatively small area burnt.
- Again, there seems to be no real trend in the area of land burnt.

A02

Application of knowledge and understanding of wildfire management, for example:

- The focus needs to be on human activities as they cause the greatest number of fires, as shown in the graphs.
- Managing this hazard can be dealt with in several ways by introducing a programme of preparedness, mitigation and prevention.
- The vegetation can be managed by creating fire breaks and using controlled burning.
- The built environment can be managed by increasing the gap between houses and vegetation and by incorporating more fire-resistant methods in construction.
- Modelling and hazard-mapping with GIS can be used to study the ways in which fires behave in order to predict fire behaviour.
- People can be educated about the risk of fires and how to avoid starting them.
- Warning systems can be put in place by establishing lookout towers, ranger patrols and air patrols, and by using local radio. In tourist areas, notice boards at strategic locations could carry warnings of the fire hazard.
- Community action programmes can be established in areas susceptible to wildfires to train local people in fire prevention and survival techniques. For example, the Community Fireguard Group in Victoria, Australia.

Hints and tips

You are asked to analyse, not simply repeat, figures from the graph. That means you have to manipulate the data and then use them to support any point you are making.

Question 6 example responses

Student A

The number of fires graph shows how fires start and it is clear that lightning plays a very small role and that most fires are caused by humans. Even in 2001, when there were the most fires caused by lightning, there were 2.5 times more human-caused fires. The average number of fires in California is around the 8000-a-year mark, with some years below this figure (2005, 2010 and 2016) and some years well above this figure (2008 and 2013). But there is no real trend except that lightning-caused fires dropped from a peak in 2001 and then remained fairly low for the next 15 years. The number of km^2 burnt graph is much more dramatic, with peaks of 5800 km^2 in 2008 (the year of the most fires) and 4600 km^2 in 2007 (another year when there were a lot of fires). In some years when there were a large number of fires, the area burnt was quite small. For example, in 2004 there were over 8000 fires but the area burnt was the fourth lowest at only 1000 km^2. This was also the case in 2011, with just under 8000 fires but the second lowest burnt area at about 550 km^2. Like the first graph, there seems to be no real trend other than some years the figures are higher than others.

As to what measures can be put in place to deal with wildfires, clearly little can be done to prevent lightning so it is the human causes that must be looked at. A comprehensive programme is needed, including how to be prepared for fires, how to mitigate the effects of fire, and then prevention itself. Forests can be managed by creating fire breaks to prevent fires from spreading and by having a programme of controlled burning of leaf litter and dead material. Creating models and hazard maps of previous fires using GIS would also help inform people in planning for the future. Educating the public about the risks is important, so that they do not inadvertently start a fire, and having patrols and lookout posts as well as notices everywhere in areas likely to be at risk. Community action is very popular in Australia, which suffers from damaging fires, and local groups get together to learn about and teach each other about fire prevention and survival techniques. This is a 'bottom-up' method and so people are more likely to take notice of it rather than a 'top-down' programme from the government.

e The student uses the data and arrives at some sensible conclusions based on the graphs. Full coverage of methods for dealing with fires follows, with evidence of synopticity, resulting in a high-level mark. High Level 2

Student B

The first graph shows that are a lot of fires caused by humans and few by lightning. The worst year for lightning was 2001 when it started 3000 fires, but since then that number has fallen to about 1000 every year. Human causes make up the rest and in 2008 that was over 10,000 fires. Most years there are about 8000 fires, give or take. The area graph is the land that is damaged, but the graph does not say that so I am guessing really. The worst year was 2008, but then that was the year they had most fires. Other years do not follow any real pattern, as a lot of fires in one year might do little damage and then fewer fires in another year might do a lot. Neither graph has a trend.

To deal with fires you need to educate people not to start fires and then what to do if they do. Lookout towers are a good idea as people can see the fire start and then warn other people or perhaps small planes could fly round and radio back if they see a fire. Perhaps a computer program could help predict where the fire will go and so make it easier to control. The forest people should make fire breaks to slow the fires down and give people a chance to get away. Small local groups could get together after they have had a fire and plan what to do next time.

e The student has a chatty style of writing that is not suitable for an A-level answer. Some points are made about the graphs, but there is much more that could be commented on. Likewise, the second paragraph lacks any real detail or structure, hence the low mark. Low Level 2/high Level 1

Question 7 mark scheme

9 marks (AO1 = 4 marks, AO2 = 5 marks). Use the levels of response mark scheme as follows.

Level 3 (7–9 marks)

- **AO1** — The answer shows detailed knowledge and understanding.
- **AO2** — Knowledge and understanding are appropriately applied. Connections and relationships are made between the different aspects studied, which are fully developed and completely relevant. The evaluation is detailed and well supported. The answer provides a well-balanced and coherent argument.

Level 2 (4–6 marks)

- **AO1** — The answer shows clear knowledge and understanding, but there may be some minor inaccuracies.
- **AO2** — Clear knowledge and understanding are appropriately applied. Connections and relationships are made between the different aspects studied, which have some relevance. The evaluation is clear and is supported with appropriate evidence. The answer provides a clear but partial argument.

Level 1 (1–3 marks)

- **AO1** — The answer shows basic knowledge and understanding, with limited relevance, and there are inaccuracies.
- **AO2** — Limited knowledge and understanding are applied. The connections and relationships made between the different aspects studied are basic, with limited relevance. The evaluation is basic. The answer provides a basic argument.

Notes for answers

AO1 Knowledge and understanding of the elements of a management plan:

- Define risk.
- Prediction to include monitoring and risk assessment.
- Prevention and mitigation.
- Preparedness.
- Post-disaster response.
- Post disaster reconstruction; economic and social recovery.
- Locational reference.

AO2 Application of knowledge and understanding to located place with associated detail:

- Relevant connections and relationships are observed and commented on.
- Evaluation of the scheme in terms of success for the located place.
- Well-balanced arguments are considered.

Hints and tips

Here is an opportunity to show wider reading and research into this topic. Making detailed use of your case study to illustrate your points is essential to gain the higher-level marks.

Question 7 example responses

Student A

The Philippines is one of the most hazard-prone countries in the world, suffering from earthquakes, volcanic eruptions, typhoons, landslides and tsunamis. A recent 7.1 magnitude earthquake killed 185 people, injured 583 and displaced 78,500 families. The average cost of these events to the country each year is about US$200 million. All of these events are largely unavoidable so the response has to be a wide-ranging, holistic approach and not just a post-event response approach.

The Philippines government set up a Disaster Risk Reduction and Management Policy (DRRMP) in 2009 to look at how the country could develop sustainably while dealing with the impact of natural disasters.

Pre-disaster measures such as educating the local population in what to do before, during and after an event has been a major part of the programme, and also realising the need for stockpiling medical equipment, tents, water-purifying equipment and food supplies. Another major part of the preparation is using hazard maps created by a GIS. This allows people in high-risk areas to be given warnings and even evacuated. The problem with both strategies is that, as an LIC, the Philippines may not be able to afford to have supplies sitting about waiting for disaster to strike. Population pressure also means that the very poor have to live in hazard-prone areas that no one else wants to live in, so this group is instantly at risk through no fault of their own.

The response part of any management plan is firstly about saving people and providing emergency assistance. Often this requires outside help from NGOs like Oxfam, MSF and the Red Cross, as countries like the Philippines do not have this type of support available. A later stage of the response plan is to assess the damage and plan for post-disaster recovery. The Philippines is used to dealing with disasters, having suffered more than 360 since 1980, and its DRRMP attempts to deal with this.

A key part of the plan is to reduce disaster vulnerability through poverty reduction and build resilience in sectors of the population that are especially vulnerable, such as the urban poor. Recovery takes varying lengths of time depending on the magnitude of the disaster, as there may need to be a programme of reconstruction following an earthquake, or economic and social recovery as in the case of the Pinatubo eruption, which caused widespread devastation of farms and farmland. At the same time, the country must continue with its normal programme of development activities.

The success of any plan can only be measured against its performance in the face of a disaster. The Philippines continues to suffer with increased numbers of typhoons and the ever-present threat of seismic and volcanic activity, and it will only be when disaster strikes that the plan can be evaluated.

e The student has written a lot in the time available and they have clearly done a lot of individual research; this is reflected in this detailed and synoptic answer. An excellent level of language is used in this well-constructed response. High Level 3

Student B

The USA has an organisation called FEMA, which was set up in the 1970s to help direct the response to disasters within the USA and where the local and state organisations cannot cope. FEMA works to a plan called the disaster management cycle, which has three stages: pre-disaster, response to the disaster and post-disaster recovery. The preparedness phase comes before the disaster hits and includes things like making hazard maps using GIS, preparing people by getting them to board up their windows before a hurricane hits, and in California storing food and medical supplies in case of an earthquake.

The next phase is immediately after the event when emergency search and rescue facilities have to be put in place to save people and provide emergency housing and medical care. An assessment has to be made to see what will be required in the last phase, which is post-disaster recovery.

This last phase could probably be the longest as it may take years for the recovery to be completed — rebuilding houses, shops and offices to get the economy going again. The social recovery may take forever in the case of people who lose loved ones in the disaster, something that will remain with them always.

The disaster plan for New Orleans and the role of FEMA following Hurricane Katrina was severely criticised as the defences were breached and then water could not get back out into the river and sea. People (mainly elderly and black people) were left stranded in the flooded parts of the city and large parts of the city remain ruined some 12 years later. So as a plan it was not a very good one.

e The student makes a reasonable effort using a located area and referencing other places within the USA. The answer lacks detail, which has restricted the mark achieved. High Level 2

Question 8 mark scheme

20 marks (AO1 = 10 marks, AO2 = 10 marks). Use the levels of response mark scheme as follows.

Level 4 (16–20 marks)

AO1

- Detailed, relevant and appropriate knowledge and understanding of place(s) and environments.
- Full and accurate knowledge and understanding of key concepts.
- Detailed awareness of scale and temporal change where appropriate.

AO2

- Detailed evaluative conclusion.
- Detailed application of knowledge and understanding throughout.
- Full evidence of links between knowledge and understanding.

Level 3 (11–15 marks)

AO1

- Generally clear and relevant knowledge and understanding of place(s) and environments.
- Generally clear and accurate knowledge and understanding of key concepts.
- Generally clear awareness of scale and temporal change where appropriate.

AO2

- Clear, evaluative conclusion.
- Generally clear application of knowledge and understanding.
- Generally clear evidence of links between knowledge and understanding.

Level 2 (6–10 marks)

AO1

- Some relevant knowledge and understanding of place(s) and environments.
- Some knowledge and understanding of key concepts.
- Some awareness of scale and temporal change where appropriate. There may be a few inaccuracies.

AO2

- Some sense of a conclusion.
- Some partially relevant application of knowledge and understanding.
- Some evidence of links between knowledge and understanding.

Level 1 (1–5 marks)

AO1

- Very limited knowledge and understanding of place(s) and environments.
- Very limited knowledge and understanding of key concepts.
- Very limited awareness of scale and temporal change where appropriate. There may be a number of inaccuracies.

AO2

- Very limited and/or unsupported conclusion.
- Very limited application of knowledge and understanding.
- Very limited evidence of links between knowledge and understanding.

Notes for answers

A01 Knowledge and understanding of landscape features produced by tectonic processes:

- The key reasons for different types of feature: type of plate boundary, faults and hot spots.
- Extrusive landforms: volcanoes, fissure eruptions, lava fields, long-term Deccan Plateau, short-term cones.
- Intrusive features: batholiths, dykes and sills.
- Earthquake activity: ground displacement, rift valleys.
- Secondary impacts of earthquakes: tsunamis remodelling coastlines, damage to coral reefs.
- Secondary features of volcanoes: lahars.

A02 Application of knowledge and understanding:

- The changes in landscapes resulting from human activity: farming of fertile land, use of geothermal energy, tourism.
- An understanding of the scale of the different features.
- Consideration of the age of the landscape, the effects of erosion.
- Extensive use of specialist language.
- Use of well-referenced examples.

Hints and tips

The quality of your answer will be determined by the way in which your points are supported by reference to a place or places that you have studied. Depth, detail and up-to-date material will also benefit your answer.

Using paragraphs will help isolate your ideas and give your answer a much better structure. Include a brief introduction, setting the scene for what follows, and an equally brief conclusion referring back to the question to round things off at the end.

Question 8 example responses

Student A

Tectonic processes can produce a variety of landforms, depending on several factors. Some landforms result from the formation of the landscape itself but others will result from secondary effects.

The plate boundaries that exist within the Earth's crust give rise to a variety of different landscapes. Destructive margins produce deep-sea trenches, mountain ranges and volcanoes, as shown down the west coast of South America. The volcano is an extrusive feature of the landscape and can take several different forms. The explosive type may produce larger features such as calderas, some of which are so big they have only recently been identified from satellite pictures, such as the supervolcano in Yellowstone National Park. When Pinatubo erupted in 1991, it blew 300 m off its summit, changing the landscape in the area forever. Less explosive fissure volcanoes may produce much larger features that extend over many hundreds of square kilometres, such as the Indian Deccan Plateau.

Where the plate boundary is a constructive margin, the features are much less obvious and consist of extensive basaltic lava fields, which we saw in fieldwork in Iceland.

Other landscapes will evolve from intrusive activity with the formation of batholiths — cooled-down plutons of magma resulting from deep melting within the crust. These may be eroded over many millions of years, for example to give the landscape of moors in southwest England. The granite tors of Dartmoor and Bodmin Moor are eroded intrusive features that make up the surface of the modern landscape. Where magma is intruded along a bedding plane, sills are formed vertically through the strata, giving rise to dykes. The Great Whin Sill runs across northern England, forming the border between Roman England and Scotland, and for much of its length the Great Whin Sill is topped by Hadrian's Wall.

Earthquake activity affects the landscape in a variety of ways, from a small landslide to a major displacement of land. The African Rift Valley, the Rhine Valley and the Eden Valley in northwest England are all the result of displacement and all have a distinctive landscape.

Earthquakes can also cause tsunamis, as seen in Japan in 2011. These powerful waves have the ability to reshape coastlines by removing or adding sediment and by damaging the coasts that are fringed by coral reefs. The tsunami is a secondary effect, as are the lahars and floods that result from volcanic activity. The slopes of Pinatubo are coated in a mud flow several metres thick, making a radical difference to the landscape.

Where tectonic activity produces resources that humans can use, such as soil and geothermal energy, the landscape will be altered. In Iceland, the growth of greenhouse cultivation based on the geothermal heat has created a very distinctive landscape. Rich volcanic soils mean that mountain slopes are heavily cultivated, producing the farming landscape seen in Italy on the slopes of Etna and Vesuvius.

The range of tectonic processes creates a similarly wide range of landscapes at different scales, which are subject to secondary effects and erosion.

e The student provides a comprehensive review of the variation of tectonic landscapes, referring where appropriate to both fieldwork and other locations. Level 4

Student B

The main tectonic processes are those associated with the movement of plate margins. Destructive margins produce a landscape full of volcanoes and prone to earthquakes, whereas constructive margins give us lava fields like in Iceland. Conservative margins, as in California, give a badly faulted landscape, but to the north volcanoes appear as one plate goes under the other. There are different types of volcano — some are upright as the lava doesn't flow, like in central France, whereas others are flat as the lava flows for miles, like the Hawaiian volcanoes. In Iceland, where there is a constructive margin, the landscapes are mainly lava fields that are pretty flat, although there are some volcanoes as well that produce lava and ash. When the volcanoes are under the ice, the lava melts the ice and the huge amounts of water alter the landscape, but this is a secondary effect so might not count.

Sometimes the tectonic processes cause molten rock to be pushed up to the ground from below. This is what caused the formation of Dartmoor and the Great Whin Sill that Hadrian's Wall is built on. These were formed by similar processes but are two very different landscapes.

Earthquakes can cause rift valleys like the Eden Valley in Cumbria, which has very steep sides as the land in the middle dropped down, leaving the steep sides. They can also cause landslides like in the Philippines and in Italy in January 2017, when a series of earthquakes caused a landslide to hit a hotel.

e The student touches on some of the aspects of tectonic landscapes, but with no real detail and without the necessary accuracy of facts and use of specialist language. The answer lacks synopticity and a conclusion. Middle Level 2

Section A Global systems and global governance

Set 1

Question 1 mark scheme

4 marks (AO1 = 4 marks). 1 mark for each valid point, with marks available for development.

Notes for answers

Globalisation has impacted flows of capital in a number of ways:

- Deregulation has led to a removal of restrictions on the movement of capital.
- There are now major flows of foreign direct investment (FDI), where a government or a large company invests in a foreign enterprise.
- Repatriation of profits — this is the return home of profits made abroad.
- Aid has increased; the flow is mainly from rich countries to poor countries.
- Remittance payments are one type of capital flow that has been significantly affected by globalisation as economic migrants seek work in countries away from their homeland. A remittance payment is the money they send to their families back home. This boosts the economy of the home country.

Question 1 example responses

Student A

Globalisation is the increased interconnectedness of the world. These connections take place in many ways, including flows of capital, which have become more flexible and widespread since deregulation.

Increased types of capital flow as the result of globalisation include remittance payments, which are monies sent to a home country by economic migrants working abroad. This money boosts the economy of the home country as it provides spending power.

Foreign direct investment is investment by a government or company in an enterprise abroad. Most of this is controlled by developed nations.

e The student covers four clear points. 4 marks

Student B

There are now more economic migrants, and remittance payments represent capital that is sent home by migrant workers.

Foreign direct investment is investment by a government or a company in an enterprise abroad.

Deregulation has allowed freedom of movement of capital.

e The student covers three points; some development or an additional point is needed to gain full marks. 3 marks

Question 2 mark scheme

6 marks (AO3 = 6 marks). Use the levels of response mark scheme as follows.

- **Level 2 (4–6 marks):** Clear analysis of the data is provided, which makes good use of the evidence to support the points. Clear connections are made between different elements of the data.
- **Level 1 (1–3 marks):** Basic analysis of the data is provided, which makes limited use of the evidence to support points. Basic or limited connections are made between different elements of the data.

Notes for answers

- The data illustrate significant increases in remittance payments in all selected countries.
- All of the payments shown have increased significantly, from about or below US$50bn to over US$650bn the case of FDI.
- Aid used to be the highest form of payment but is now the lowest. It has only increased by about US$60bn. It is the only one not to fall during the credit crunch.
- The biggest rise is in FDI, which peaked in 2007 just before the world financial crisis. During this, it fell by US$200bn before rising again.
- Remittances have shown a steady rise, with only a small dip around 2008–09.
- Private debt increased rapidly during the early part of the twenty-first century and then fell the most as a result of the global financial crisis.

Hints and tips

Analysing data requires you to break down the findings into different parts and then make logical connections using evidence to show the relationship between the different parts.

Question 2 example responses

Student A

All the selected payments grew between 1990 and 2012. Aid, the highest amount in 1990, has doubled in size but became the smallest flow in 2003 and has continued being the lowest. FDI, the second lowest amount in 2009, has increased by over 60 times and is now the largest flow. Both FDI and private debt increased rapidly during the early part of the twenty-first century and they both showed a massive dip as a result of the global financial crisis, with private debt falling the most. Although it partly recovered by 2010, it has fallen since. The flow least affected in 2007 is remittances. The line got steeper towards 2008 followed by a small dip. Now it continues to rise.

The student shows clear analysis, correctly using the data. There are connections made and supporting evidence quoted. **Top Level 2**

Student B

All of the flows of capital have risen, with FDI being the one that has risen the most. The one that has risen the least is ODA, which rose about US$50bn in the 21 years. They all have a dip in them at about the same time in the late noughties, except for ODA.

ⓔ **The student has described the data, but not made any analysis.** **Top Level 1**

Question 3 mark scheme

6 marks (AO1 = 2 marks, AO2 = 4 marks). Use the levels of response mark scheme as follows.

Level 2 (4–6 marks)

- **AO1** — Demonstrates clear knowledge and understanding of concepts, processes, interactions and change.
- **AO2** — Applies knowledge and understanding to the novel situation, offering clear analysis and evaluation drawn appropriately from the context provided. Connections and relationships between different aspects of study are evident, with clear relevance.

Level 1 (1–3 marks)

- **AO1** — Demonstrates basic knowledge and understanding of concepts, processes, interactions and change.
- **AO2** — Applies limited knowledge and understanding to the novel situation, offering basic analysis and evaluation drawn from the context provided. Connections and relationships between different aspects of study are basic, with limited relevance.

Notes for answers

AO1 Knowledge and understanding of the changes to Antarctic sea ice:

- The sea ice is decreasing, but the retreat is not linear. Some years are worse than others.
- The retreat of the sea ice is not uniform across the continent.
- In east Antarctica the sea ice is expanding towards the Indian and Pacific Oceans.

AO2 The data show that:

- In the ten years between 2008 and 2018, the sea ice has decreased through all of the months.
- The greatest decrease is between December and May, summer/autumn in Antarctica.
- The least decrease is between June and October, the southern winter and spring.

Hints and tips

The examiner has given you some data, but simple repetition or analysis will not get you beyond Level 1. You need to use the data and add some knowledge of your own. More complex analysis could, for example, involve looking at the seasonality of the changes. Further helpful information about patterns of melting or changes in the ice cap will move your answer to Level 2. A final way would be to indicate whether there are any feedback loops that may affect future melting.

Question 3 example responses

Student A

The table shows that there was a lot less sea ice in 2018 than there was in 2008, with a huge loss of over 2.5 million km^2 in January. The differences between the amounts in the two years changes throughout the year, with the greatest difference in January. This is right in the middle of summer in Antarctica.

From my own studies I know that the decrease is not steady. 2013 showed an increase in sea ice, though it fell again the next year. Also, the change is not uniform across the continent, with some areas like the east of Antarctica growing while other parts are shrinking. This is because the winds and ocean currents are changing. Since 2013 the rate of ice melting has accelerated and scientists are worried that it will get worse.

e The student has used the table to assess the extent of the changes and has also brought in their own knowledge. **Level 2: 6 marks**

Student B

There is less sea ice in Antarctica in 2018 than in 2008. This is due to global warming, where carbon in the atmosphere is heating it up and this melts the ice. The amount it has decreased is not the same through the year. It is greatest in January and least in July. The ice is growing in eastern Antarctica.

e The student offers basic description of the change in sea ice, and adds some basic information of their own. There is no attempt at assessment. **Level 1: 2 marks**

Question 4 mark scheme

20 marks (AO1 = 10 marks, AO2 = 10 marks). Use the levels of response mark scheme as follows.

Level 4 (16–20 marks)

AO1

- Detailed, relevant and appropriate knowledge and understanding of place(s) and environments.
- Full and accurate knowledge and understanding of key concepts.
- Detailed awareness of scale and temporal change where appropriate.

AO2

- Detailed, evaluative conclusion.
- Detailed application of knowledge and understanding throughout.
- Full evidence of links between knowledge and understanding.

Level 3 (11–15 marks)

AO1

- Generally clear and relevant knowledge and understanding of place(s) and environments.
- Generally clear and accurate knowledge and understanding of key concepts.
- Generally clear awareness of scale and temporal change where appropriate.

AO2

- Clear, evaluative conclusion.
- Generally clear application of knowledge and understanding.
- Generally clear evidence of links between knowledge and understanding.

Level 2 (6–10 marks)

AO1

- Some relevant knowledge and understanding of place(s) and environments.
- Some knowledge and understanding of key concepts.
- Some awareness of scale and temporal change where appropriate. There may be a few inaccuracies.

AO2

- Some sense of a conclusion.
- Some partially relevant application of knowledge and understanding.
- Some evidence of links between knowledge and understanding.

Level 1 (1–5 marks)

AO1

- Very limited knowledge and understanding of place(s) and environments.
- Very limited knowledge and understanding of key concepts.
- Very limited awareness of scale and temporal change where appropriate. There may be a number of inaccuracies.

AO2

- Very limited and/or unsupported conclusion.
- Very limited application of knowledge and understanding.
- Very limited evidence of links between knowledge and understanding.

Notes for answers

For the host country, benefits include the following:

- Jobs and income are created — link to the multiplier effect.
- New technology is brought in.
- Workers are provided with new skills.
- Tax revenue is increased — link to the multiplier effect.
- Higher-level answers may pick up on the tax incentives offered by host countries to attract TNCs, e.g. Apple and the Irish government.

For the host country, problems centre on working practices and environmental issues:

- Poor working conditions.
- Exploitation of resources.
- Pollution and waste, which have negative environmental impacts.
- The globalising effect of TNCs, which can impact on local cultures.
- Repatriation of profits — economic leakage.
- Top-level jobs are retained by workers from the country of origin.
- Branch plant closure — branch plants are often the first to close if cheaper sites are found or if there are changes in the host country's economy, e.g. rising wage levels.
- A lack of decision-making at the local level and a lack of control by the host government.

For the country of origin, benefits include the following:

- Cheap labour.
- Cheaper goods.
- More profits.
- Access to new markets.
- Polluting manufacturing plants go overseas — possible environmental benefit.
- Allows for the specialisation of research and development and financial services elsewhere.

For the country of origin, problems include the following:

- Loss of manufacturing jobs overseas — deindustrialisation.
- Structural unemployment.
- Investing overseas at a time of recession is not a popular policy.

The conclusion may address the fact that it is hard to put a monetary value on many of the issues discussed, e.g. the impact of TNCs on the host culture or pollution issues.

Hints and tips

Whenever you start a 'To what extent' answer, it is always best to state your viewpoint straightaway. Then spend the rest of the time justifying that viewpoint. Your essay should be based on an analysis and evaluation of the facts. Evaluation means looking at an issue from different sides, covering its strengths and weaknesses. There is a lot of information to cover, so a clear structure is important. You could divide your essay into four sections, discussing the benefits and problems as set out in the notes above. Using a detailed case study to illustrate your points is essential to gain the higher-level marks. Always finish with a clear, summative statement that brings together all your arguments.

Question 4 example responses

Student A

Transnational corporations (TNCs) are companies that operate in at least two countries. Usually there is a headquarters based in the country of origin and branch plants in a number of other countries. TNCs represent many different sectors, for example manufacturing, tourism, resource extraction and call centres. They may originate from developed or emerging economies. TNCs choose to locate different parts of their operations in different countries to escape trade tariffs (e.g. Nissan car production in Sunderland), to find the lowest-cost location for their production (especially labour costs), to reach foreign markets more effectively (e.g. McDonald's) or to make use of resources that may be difficult to transport from foreign countries. TNCs can benefit both their host countries and countries of origin, but problems for both can also arise. This essay will show that the costs and benefits are wide-ranging and difficult to compare directly as many cannot be quantified in monetary terms.

The benefits to the host country mainly include the advantages associated with job creation, which are fed through the economic system of the host through the multiplier effect. Having people in waged employment means that more people pay taxes, which can be used to finance the host country's development. People also have money to spend, which boosts the economy, and they may potentially have money to save, which boosts the ability of the banks to lend. There are additional advantages of workers being trained and acquiring new skills. For the country of origin, the benefits include cutting costs of production — mainly regarding labour costs — which maximises profits for the company.

The problems with TNCs include a lack of control for the host country over decision-making and the constant concern that branch plants may close if cheaper locations emerge. Poor working conditions and excessively low wages have also been a constant criticism of TNC practices.

Apple is a US-based electronic technology company. It has 98,000 employees, over 450 retail stores in 16 countries and it is the eleventh largest TNC in the world. Its main offices and R&D are based in California and North Carolina in the USA. There is a European headquarters in Cork, Ireland, and manufacturing plants in China and Hong Kong. The main advantage to Apple of having manufacturing plants in China is a large, well-skilled, hard-working and low-cost workforce. However, it has been claimed that 200,000 workers are kept in 'sweatshop' conditions, working long hours for excessively low pay. It was also proved that a toxic chemical was used to clean iPad screens in these plants. In America, there are high levels of unemployment and in times of recession the loss of manufacturing jobs to China is not popular.

Apple employs 4000 workers in Ireland, but many of the highest-paid jobs go to foreign nationals. There has been controversy over tax issues, with claims that the Irish government offered favourable treatment on tax in return for Apple's investment.

In conclusion, where TNCs locate their activities can bring both costs and benefits to the host countries and to the countries of origin. It is difficult to balance these in monetary terms; pollution and the loss of culture are impacts in host countries that cannot be quantified. At the end of the day, the balance is complex and depends on the exact nature of the business and the agreement with the host.

e **The student provides a well-balanced argument, with a clear structure, supporting evidence from a case study and a clear conclusion. The introduction could be shorter, as the answer is a bit too long.** **Level 4**

Student B

TNCs are large multinational companies. They have a head office in their country of origin and set up branch plants in other countries where they can minimise costs and maximise profits. For many companies this means making the most of cheap labour costs in the Far East, where standards of living are lower. However, there are advantages and disadvantages.

The advantages to a host country of having a TNC such as Apple locate a manufacturing plant include the fact that unemployment in the host country will fall. People who are in waged employment are able to pay taxes. The TNC being from a developed part of the world will also bring in new technology and new ideas. The workers in the factories will learn new skills that can be transferred to other manufacturing plants in the host country and they may be able to build up a specialist workforce. This is all called the multiplier effect.

Problems include the fact that the factories are often overcrowded and poor in terms of facilities and that the wages paid are very low. Workers are sometimes exploited in sweatshops. If the factory has a lot of waste or is polluting the atmosphere, the effects will be felt in the host country. Apple used a toxic chemical to clean the screens of its iPads and this affected workers in the factory in China. In the same factory, workers were paid just $100 a month for 60 hours' work a week.

On the other hand, Ikea has a positive impact on the countries it locates in. It has strict rules on wages and conditions for workers in the countries in which it locates and it has quotas for sustainable timber products. This is an example where the impacts are monitored. Many TNCs in the fashion retail industry do not have a good impact. In a sports shoe factory in Jakarta, wages are just US$0.42 per hour and not enough to meet basic living costs. Overtime is compulsory and illegally high — workers often work from 7 a.m. to 9 p.m. and may work for a month without a day off. Workers can also be sacked for being in a union.

In conclusion, TNCs provide many benefits but there are also costs involved, which mainly impact on the workers.

e The student shows some knowledge of the costs and benefits of TNCs. The multiplier effect is not clearly explained. Examples are referred to with some specific details and they do highlight the positive and negative ways in which the activities of TNCs are conducted. In general, more depth and detail are required for the higher Level 4 marks. Level 3

Set 2

Question 5 mark scheme

4 marks (AO1 = 4 marks). 1 mark for each valid point, with marks available for development.

Notes for answers

The answer could cover the IMF or the World Bank or any other relevant international organisation:

- The IMF is a global regulator of financial flows; the World Bank aims to support less developed countries and reduce poverty.
- Both organisations provide loans, but the IMF is seen as a 'lender of last resort' and will make loans only to prevent a global crisis.
- The World Bank provides interest-free loans to low-income countries.
- The IMF provides financial help with balance of payments to countries in need.
- The World Bank is a large organisation with 7000 staff and 185 member countries; the IMF has 2300 staff and 185 member countries.
- The main role of the World Bank is to promote economic development in low-income countries.
- Both organisations have been criticised in their respective tasks.

Question 5 example responses

Student A

The World Bank has evolved into its present-day role. In its early formation it had a focus on countries in the developing world — the World Bank increased the size and number of loans to such borrowers because its main role was to enable development to take place and poverty to be reduced.

There was also a policy move towards building schools and hospitals, improving literacy and agricultural reform, as these were seen as projects that helped economic and social development.

By the late 1990s there was a growing need to help poorer countries service their debts because many countries could not afford the interest payments and this was hindering development.

From the 1990s onwards the World Bank had a policy towards achieving the Millennium Development Goals.

e The student provides four developed points about the role of the World Bank. 4 marks

Student B

I have chosen the International Monetary Fund (IMF). It:

- oversees the global financial system
- offers financial and technical assistance to its members
- provides loans only if it will prevent a global economic crisis — it is called the international 'lender of last resort'
- draws its financial resources from the quota subscriptions of member countries.

e The student lists points that describe the role of the IMF. Lists are not ideal. All exam answers should be in prose. 3 marks

Question 6 mark scheme

6 marks (AO3 = 6 marks). Use the levels of response mark scheme as follows.

- **Level 2 (4–6 marks):** Clear analysis of the data is provided, which makes good use of the evidence to support the points. Clear connections are made between different elements of the data.
- **Level 1 (1–3 marks):** Basic analysis of the data is provided, which makes limited use of the evidence to support points. Basic or limited connections are made between different elements of the data.

Notes for answers

- The overall global pattern is extremely complex and shows that some regions are distinctly areas of supply — the Middle East and the former Soviet Union — and others regions are key areas for demand — China, Europe and North America.
- The Middle East exports most of its oil to Asia (over 10 million barrels per day). Much smaller amounts go to Europe (just over 1.7 million barrels per day) and North America (2.1 million barrels per day). Small amounts go to Africa and South America (approximately 0.5 million barrels per day combined).
- China and Asia are major areas of demand. Asia has a diverse supply — it receives oil from the Middle East (10.6 million barrels per day combined), China, Africa, Europe and Latin America. China also receives oil from Asia, Europe, the Middle East, Africa and Latin America (over 2.5 million barrels per day in total).
- North America is also a major area for demand (over 7 million barrels per day). It is the second largest destination region.
- Europe and Latin America have flows into and out of the region. For example, Europe as a destination receives approximately 9 million barrels per day, making it the largest destination region, with most of this coming from the former Soviet Union (4.7 million barrels per day).
- The stand-out supply regions are the Middle East (approximately 15 million barrels per day) and the former Soviet Union (approximately 6 million barrels per day), with net outflows that illustrate the geopolitical influence of these regions.

Hints and tips

Analysing data requires you to break down the findings into different parts and then make logical connections using evidence to show the relationship between the different parts.

Take care with the units — the map shows units of 1000 barrels per day, so 30 is 30,000 barrels per day and 1000 is 1,000,000, i.e. 1 million barrels per day.

Question 6 example responses

Student A

The map shows a complex pattern. The main areas that export oil are the Middle East and the former Soviet Union. The main customers are Asia for the Middle East (over 10 million barrels per day) and Europe for the former Soviet Union (over 4 million barrels per day). The main link is between the Middle East and Asia; it is over twice the volume of any other flow.

The main receiving areas are Europe (almost 10 million barrels per day) and North America (approximately 7 million barrels per day).

Africa has inflows and outflows. The outflows are mainly to Europe, of more than 2 million barrels per day.

Latin America has a small amount of inflows (just over 0.5 million barrels per day) and major outflows to North America (2.6 million barrels per day).

The map says a lot about the geopolitical ties of regions, e.g. the former Soviet Union and Europe, and the Middle East and Asia.

e **The student provides a structured account of the high flows for the destination and supply areas, and all the regions are commented on. Comparative comments are also made and supported with accurate evidence.** **High Level 2**

Student B

The Middle East exports the most crude oil globally and this is mainly to Asia (10,661 thousand barrels per day). North America and Europe also are major destination areas; outside of the Middle East, North America also receives crude oil from South America.

Africa is not completely marginalised in the pattern — it is a destination area for four other regions and it supplies Europe, China, Asia and North and South America.

Europe has a strong link with the former Soviet Union (4,764,000 barrels per day).

e **The student gives an overview of the main trends, but supporting evidence is lacking for most of the statements and the coverage is not detailed.** **Low Level 2**

Question 7 mark scheme

6 marks (AO2 = 4 marks; AO3 = 2 marks). Use the levels of response mark scheme as follows.

Level 2 (4–6 marks)

- **AO2** — Applies knowledge and understanding to the novel situation, offering clear analysis and evaluation drawn appropriately from the context provided. Connections and relationships between different aspects of study are evident, with clear relevance.
- **AO3** — Clear analysis of the quantitative evidence provided, which makes appropriate use of evidence in support. Clear connection(s) between different aspects of the evidence.

Level 1 (1–3 marks)

- **AO2** — Applies limited knowledge and understanding to the novel situation, offering basic analysis and evaluation drawn from the context provided. Connections and relationships between different aspects of study are basic, with limited relevance.
- **AO3** — Basic analysis of the quantitative evidence provided, which makes limited use of evidence in support. Basic connection(s) between different aspects of the evidence.

Notes for answers

AO3

- There were no visitors to Antarctica before 1984–85, apart from small numbers in 1966–67 and 1976–79.
- Since 1984–85 the numbers have increased, to a maximum of 580 in 2016–17.
- The rise has not been steady, with falls in 1994–95, for example, when numbers were zero.

AO2 Application of knowledge and understanding to analyse and evaluate the extent to which Antarctic is vulnerable to visitors. Recovery from damage to ecosystems is slow and uncertain. For example:

- The harsh land conditions support a relatively small number of animal species that are very carefully adapted to the conditions. Some animals are ashore during the breeding season and are easily disturbed because this is the tourist season.
- The Antarctic ecosystem is extremely fragile — disturbances leave their imprint for a long time, e.g. footprints on moss can last for decades.

Hints and tips

Link the increasing numbers of visitors to the increase in potential damage caused by these visitors.

Question 7 example responses

Student A

Although the graph shows that tourist numbers have increasd from virtually zero before 1985 to around 600 in 2016, the numbers are still very low because this is a dangerous place to be. Those who do go ashore are mainly interested in the Antarctic wildlife. The very short time that it is safe for people to visit ashore is also the height of the breeding season for the animals they have come to see, e.g. seals and penguins. The boats that tourists arrive on pollute the air with exhaust fumes, but this is really quite a small problem. The big problem is when these ships bring with them exotic species attached to the hull of the boat. These have a massive effect on the marine animals.

Plant species are also delicate and human footprints on mosses can last for decades. Although harsh, the environment is extremely fragile.

e The student shows clear and accurate knowledge and understanding of the Antarctic environment. Connections are made between the increased numbers of visitors and the effect they have on the environment. **Level 2**

Student B

Antarctica is a very fragile environment. There are very small food chains and food webs with very low biodiversity. Therefore, if just one species suffers or even becomes extinct due to an increase in visitor numbers, this will have a major impact on the rest of the ecosystem.

The increase in land-based visitors during the summer time means that animals like seals are disturbed when breeding and the species may become extinct.

All those visitors trampling about leave their imprint for a long time, e.g. footprints on moss can last for decades.

e The student shows some knowledge and understanding of the impact of visitors but has made little reference to the graph. Level 1

Question 8 mark scheme

20 marks (AO1 = 10 marks, AO2 = 10 marks). Use the levels of response mark scheme as follows.

Level 4 (16–20 marks)

AO1

- Detailed, relevant and appropriate knowledge and understanding of place(s) and environments.
- Full and accurate knowledge and understanding of key concepts.
- Detailed awareness of scale and temporal change where appropriate.

AO2

- Detailed, evaluative conclusion.
- Detailed application of knowledge and understanding throughout.
- Full evidence of links between knowledge and understanding.

Level 3 (11–15 marks)

AO1

- Generally clear and relevant knowledge and understanding of place(s) and environments.
- Generally clear and accurate knowledge and understanding of key concepts.
- Generally clear awareness of scale and temporal change where appropriate.

AO2

- Clear, evaluative conclusion.
- Generally clear application of knowledge and understanding.
- Generally clear evidence of links between knowledge and understanding.

Level 2 (6–10 marks)

AO1

- Some relevant knowledge and understanding of place(s) and environments.
- Some knowledge and understanding of key concepts.
- Some awareness of scale and temporal change where appropriate. There may be a few inaccuracies.

A02

- Some sense of a conclusion.
- Some partially relevant application of knowledge and understanding.
- Some evidence of links between knowledge and understanding.

Level 1 (1–5 marks)

A01

- Very limited knowledge and understanding of place(s) and environments.
- Very limited knowledge and understanding of key concepts.
- Very limited awareness of scale and temporal change where appropriate. There may be a number of inaccuracies.

A02

- Very limited and/or unsupported conclusion.
- Very limited application of knowledge and understanding.
- Very limited evidence of links between knowledge and understanding.

Notes for answers

There is a wide scope of material for this essay, which is essentially an evaluation of the impacts of globalisation on different groups, countries and societies.

A01 Knowledge and understanding of the process of globalisation and the impacts it can have.

Globalisation is a process that reflects the increased interconnectedness of the world. A global network of communications, transport, flows and trade has enabled the world's economies, societies and cultures to become more integrated.

Positive impacts may include:

- Increased economic prosperity and opportunity, particularly, some would argue, in developing nations.
- A wider allocation of resources.
- The expansion of world trade.
- The footloose nature of TNCs opening up opportunities for employment and growth in emerging economies.
- The diffusion of culture through media, television and the internet.
- The organisation of individual nations into trading blocs.
- The broadening of the global economy spatially with the emergence of a number of world cities.
- The growth of international migration spreading multiculturalism, language sharing and economic opportunities.

Impacts considered to be unfair and unequal:

- In many poorer nations, investment by TNCs is seen as a means of exploiting workers and resources in ways that benefit only the TNCs.
- The concerns over 'electronic sweatshops' due to the outsourcing of call centres and customer service work.
- Opportunities in the developed/richer parts of the world have meant that the most skilled and educated have moved away from poorer areas, hindering their development.

- There is concern in all countries regarding the erosion of their cultures. 'Nationalism' is increasingly seen as an alternative.
- The environmental impacts of industrial development, such as pollution of the air, oceans and fresh water systems, can be carried across national boundaries.

AO2 Application of knowledge and understanding to evaluate the impacts. For example:

- Creating international groupings can increase control over trade, reduce the ability of poorer nations to compete, impose unfair tariffs and increase protectionist policies.
- TNCs have a range of benefits but also costs; the exact balance and the extent to which a host country is affected depends on the context, but TNCs can increase inequality.
- Globalisation is an uneven process, e.g. the spread of telecommunications and ICT have been beneficial to many but have also left the vast numbers of people who are not 'connected' at a disadvantage.
- The outcomes of globalisation vary across the world. Many groups and societies are unaffected and marginalised, particularly by the economic benefits of globalisation. Take, for example, women, who make up 70% of the people in poverty.
- There are also differences within countries and within urban areas between those who have benefited from globalisation and those who have not.
- The increased flows of capital and labour have meant that companies can shift locations relatively easily. Wealthy companies can invest in some places and remove investment from others. The impacts of this create inequality and leave poor countries in the hands of wealthy TNCs.
- Conclusions may cover the different theoretical views on globalisation, alternatives to globalisation and the rise of nationalism.

Hints and tips

Your essay should be based on an analysis and evaluation of the facts. Evaluation means looking at an issue from different sides, covering its strengths and weaknesses.

There is a lot of information to cover, so a clear structure is important. You could divide your essay into the costs and benefits for different groups and countries, or the socio-economic, cultural, political and environmental costs and benefits.

Using examples throughout to illustrate your points is essential to gain the higher-level marks.

Question 8 example responses

Student A

Globalisation is the process by which there is a growing integration and interdependence of people's lives. Globalisation has socio-economic, cultural, political and environmental components. Its impacts have been felt across all of these dimensions but this essay will argue that the impacts are not always even and have in some cases increased inequality.

The economic impacts of globalisation centre on the emergence of a new global financial system that generates flows of capital and also enables the wider movement of people and goods. This has resulted in a global shift in the production of goods and services and the emergence of new centres of production, particularly in the emerging economies of China and India. Transnational companies such as Apple, Nokia, Toyota and Mondelēz have been able to shift production sites to maximise their profits and minimise their costs, particularly in terms of labour. Although these investments in host countries can have many economic benefits that can aid a country's development through the multiplier effect, they can also have costs that undermine a country's control and decision-making and can exploit its workers and resources. In this way inequalities are created and enhanced, not reduced.

There have also been vast technological advances through globalisation, particularly in transport due to containerisation but also in telecommunications and ICT. However, this has left those who are not 'connected' marginalised and disadvantaged. This includes mainly rural dwellers in poorer countries and people who are already disadvantaged and marginalised through poverty. This situation can create inequalities between rural and urban areas and between different groups in a country.

Globalisation has encouraged multinational migration and the free movement of people. As a knock-on effect, this has led to a multicultural society in many countries where people have benefited from learning new languages and experiencing new foods and cultures. Television, cinema and the internet have enabled this sharing. Some people are turning against this effect, however, as they fear their own native culture is being eroded; in this way, nationalism develops. The Brexit vote saw many people wanting to restore Britain's national identity and ensure that decision-making stays with the national government and not a political union. More countries are organising themselves into trading alliances but, as well as potentially reducing the power of the nation state, this is also furthering inequality as those who are not included are marginalised and denied access to certain markets, e.g. limited access to the rich markets of Europe. The NAFTA trade agreement joining North America and Mexico is an example of an agreement to join rich and poor countries.

An environmental impact of globalisation is the air, water and ocean pollution created by the industrialisation process. Here, too, impacts can spread disadvantages and it may be the case that pollution created by one country impacts another country as the effects spread in the air or water supply. Again, there will be those who gain by getting rid of their pollution and those who lose out as they have to manage it.

In conclusion, it can be seen that globalisation has a range of impacts — socio-economic, environmental, political and cultural. Some of these are positive and bring countries together in a more inclusive way, while others have a very uneven impact and continue to exclude and marginalise groups. The situation is very complex but one clear starting point would be to make trade more inclusive for poorer nations and to manage and monitor the impacts of TNCs.

e The student provides a detailed account of the impacts of globalisation, structured into the socio-economic, political, cultural and environmental impacts. Each type of impact is evaluated in terms of evenness and fairness. More examples are needed to illustrate the points, but a decisive conclusion reflects the original question. Level 4

Student B

Globalisation is the increased interconnectedness of the world. This brings many impacts, some of which seem unfair, although some impacts are good. This essay will look at examples of both and reach a concluding statement.

World trade has expanded rapidly and, although it is meant to be controlled and regulated by the World Trade Organization, in reality trade is still dominated by the advanced and emerging economies and by the affluent markets in Europe and North America. TNCs are able to locate anywhere in the world and their decisions will influence economic growth and prosperity and give the countries in which they choose to locate an unfair advantage in terms of a boost to economic growth. In this way, they have more of an influence than a country's own government.

Globalisation has also increased the free movement of people, which has had many impacts. On the positive side, it has enabled the sharing of cultures, languages and customs, and on the downside some countries have so many migrants from other countries that they fear their own identity is being eroded and so they seek policies to close their borders. Some TNCs such as McDonald's have spread a Western-style culture across the world and this is not a good thing in some countries.

The environment has also suffered. There is more production as countries develop and this creates large amounts of pollution, e.g. in China and India. For many emerging economies the pollution is not controlled and it spreads widely. This is unfair for countries that receive the impact of the pollution but did not create the pollution. Also, as more resources, goods and fuels (e.g. oil) are transported, there can be accidents such as oil slicks that pollute areas that have nothing to do with the transportation of oil.

However, globalisation has also created a lot of new opportunities as markets have developed and ideas and labour are shared. So, in conclusion, there are impacts that have been unfair and uneven but there are also many benefits for the world to come together.

e The student writes in a clear style and the essay is structured and shows a generally clear understanding and application of knowledge. More understanding of places needs to be shown through examples and the conclusion needs more depth and detail. The essay would benefit from a clear, summative conclusion. **Low Level 3**

Set 3

Question 9 mark scheme

4 marks (AO1 = 4 marks). 1 mark for each valid point, with marks available for development.

Notes for answers

- Supra-national spaces.
- Beyond the control of any nation state.
- High Seas, atmosphere, Antarctica and outer space; possibly cyberspace.
- Principle of the common heritage of mankind.
- Protected by international laws and treaties: UNCLOS, UNFCCC, ATS, Lindblad Plan, Madrid Protocol, Moon Treaty.

Question 9 example responses

Student A

The global commons refers to those areas that are beyond the reach of any country's jurisdiction. It usually applies to areas where there are shared resources, such as the sea, the atmosphere, outer space and Antarctica. Each of these areas should be available to all people for their use or for the use of future generations. Each of these common spaces is protected by international laws such as the UN Convention of the Sea (UNCLOS), the Antarctic Treaty System (ATS) and the Madrid Protocol.

e The student covers four clear points. **4 marks**

Student B

The global commons are areas that are not controlled by any one country, like Antarctica. They are protected by laws and treaties like the Antarctica Treaty and the Laws of the Sea.

ⓔ **The student covers three points; some development or an additional point is needed to gain full marks. 3 marks**

Question 10 mark scheme

6 marks (AO1 = 2 marks, AO2 = 4 marks). Use the levels of response mark scheme as follows.

Level 2 (4–6 marks)

- **AO1** — Demonstrates clear knowledge and understanding of concepts, processes, interactions and change.
- **AO2** — Applies knowledge and understanding to the novel situation, offering clear analysis and evaluation drawn appropriately from the context provided. Connections and relationships between different aspects of study are evident, with clear relevance.

Level 1 (1–3 marks)

- **AO1** — Demonstrates basic knowledge and understanding of concepts, processes, interactions, change.
- **AO2** — Applies limited knowledge and understanding to the novel situation, offering basic analysis and evaluation drawn from the context provided. Connections and relationships between different aspects of study are basic, with limited relevance.

Notes for answers

A01

- Inequality in global trade — stronger partner takes what it wants, weaker partner is in no position to resist.
- Demand in some countries exploits weaker countries for resources.
- Example — food from the Sudan to China; oil and other minerals from Chad, Kenya, Brazil.
- Bilateral agreements.

A02

- Understanding of the imbalances that exist.
- Principle of *cui bono*.
- What happens when the resources are exhausted, or more accessible resources are located?
- Issues relating to the lack of benefit actually generated.

Question 10 example responses

Student A

The diagram shows China taking over the African continent. It demonstrates the power that the Chinese have in exploiting other countries for their own benefit. The Chinese have a huge demand for resources to supply their economy and also to feed their huge population. They have the financial resources to invest in developing countries such as those in Africa to exploit the raw materials that those countries have, be it iron ore, oil or food. The Chinese can also show how beneficial their system of government is and how it has been responsible for the growth that the Chinese have experienced in the last few decades.

A bilateral agreement would be set up whereby the Chinese would provide investment for training, education and healthcare facilities to assist in the development of infrastructure to exploit the resources, while the recipient country would provide the raw materials for export. The financial investment in transportation and port facilities is something that the poorer countries do not have so they are unable to exploit their own raw materials. Once the resources become exhausted, as they must do, the Chinese will look elsewhere and the African country may be left with nothing other than some extensive port systems and social infrastructures that it can no longer afford to operate. This situation leads to greater inequality between countries and can be seen as the negative side of geopolitics.

e The student shows understanding of the disadvantages of international trade between two unequal partners. There is competent use of technical language. The student gives lots of detail and demonstrates analysis in the conclusion. 5 marks

Student B

The image shows Africa becoming part of China and supplying raw materials for its huge population and massive economy. The Chinese are very powerful players in the world trade market and everyone wants to be a partner but not everyone benefits. In particular China needs oil, which it is getting from the countries in the central part of Africa, countries that are very poor and under-developed and so keen to get Chinese money.

These African countries may benefit in the short term and get new roads and other transport systems. Their workers may get some training and there will be new schools and hospitals built as part of a bilateral agreement with China. However, when the resources run out and the Chinese leave, the African country may be worse off as it doesn't have the industry to use the transport networks and doesn't have the skilled workers to run the hospitals.

e The student gives a simplistic explanation of the situation portrayed by the image. Less attention to the image, which is only there as a stimulus, and more detail about the message behind it would benefit this response. 3 marks

Question 11 mark scheme

6 marks (AO1 = 2 marks, AO2 = 4 marks). Use the levels of response mark scheme as follows.

Level 2 (4–6 marks)

- **AO1** — Demonstrates clear knowledge and understanding of concepts, processes, interactions and change.
- **AO2** — Applies knowledge and understanding to the novel situation, offering clear analysis and evaluation drawn appropriately from the context provided. Connections and relationships between different aspects of study are evident, with clear relevance.

Level 1 (1–3 marks)

- **AO1** — Demonstrates basic knowledge and understanding of concepts, processes, interactions and change.
- **AO2** — Applies limited knowledge and understanding to the novel situation, offering basic analysis and evaluation drawn from the context provided. Connections and relationships between different aspects of study are basic, with limited relevance.

Notes for answers

AO1 Knowledge and understanding of the Antarctic Treaty:

- ATS, Madrid Protocol, Conventions on Seals, Marine Living Resources, IWC and the moratorium, Lindblad Plan, IAATO.
- The work of the above.

AO2 Application of knowledge and understanding to analyse and evaluate:

- The impact of the Antarctic Treaty, etc.
- The level of consensus among signatory nations.
- Issues affecting monitoring and reporting; environmental audits introduced by Madrid Protocol.
- Problems of enforcement, especially of overfishing and whaling.
- Growth in tourism and resultant pressures.

Hints and tips

The key to accessing the four AO2 marks is in addressing effectiveness and reaching a judgement as to how successful treaties have been.

Question 11 example responses

Student A

The Antarctic Treaty System was established in 1959 and provided the basis for the protection of this region and its surrounding seas. In addition, there is the Madrid Protocol (1991) and two conservation treaties dealing with seals and with marine resources. There was also in the 1990s a pause in whaling in this region as a result of the work of the International Whaling Commission.

The treaty system has been effective up until now and has been recognised as one of the most successful international treaties ever signed. The countries that signed up have obeyed the treaty and have observed its restrictions, which has allowed other steps to be taken to protect the plants, animals and historic sites, and also to manage tourism and allow cooperation between countries.

The Madrid Protocol was designed to prevent mining in the region and, despite some disagreement, came into force in 1998. It forbids any mining and provides for a system of monitoring and reporting, designating Antarctica as a region devoted to peace and science. Any modifications to the protocol must have unanimous agreement and nothing is likely to change until the next conference in 50 years.

Control of whaling has been less successful as the Japanese continue to hunt whales despite objections from many other countries. Given the difficult conditions, this is an activity that is hard to police. More of an issue is the question of overfishing in the seas around Antarctica. As more and more fish are taken some species are being seriously depleted, giving rise to concern. Many of the species are low on the food chain and as a result other, higher-order consumers are also put at risk.

Tourism has been strictly controlled by rules in the Lindblad Plan, which manages tourists to have the lowest impact possible, especially where they might come into contact with land species that have very few predators and are therefore at a higher risk through their inquisitiveness.

In conclusion, this is one area where international organisations have been very successful in managing the vulnerable species and areas that make up Antarctica.

e The student covers a range of content, showing good knowledge and understanding. There is some application of this knowledge and assessment. 6 marks

Student B

The Antarctic Treaty consists of the 12 countries that signed the ATS in 1959. This prevents Antarctica being used for anything other than scientific research. This has been very successful and other laws to protect the species have been introduced, for example the Madrid Protocol. This prevents mining, which could have a bad effect on the penguins and other animals. Fishing and whaling are also limited by the International Whaling Commission, which allows only certain species to be caught at certain times of the year. This has not been very successful and overfishing occurs. It is difficult to prove if someone is overfishing as there are not enough patrol boats. Tourists are a big problem for the animals that live here and the numbers of tourists allowed to land is strictly controlled, so this has been quite successful.

Apart from the fishing and the whaling, the agreements have been fairly successful in protecting the species in this area — better than elsewhere.

e The student demonstrates basic knowledge and understanding of some of the international agreements at work in this area. There is limited discussion on the wider threats to land animals, whaling and overfishing. 3 marks

Question 12 mark scheme

20 marks (AO1 = 10 marks, AO2 = 10 marks). Use the levels of response mark scheme as follows.

Level 4 (16–20 marks)

AO1

- Detailed, relevant and appropriate knowledge and understanding of place(s) and environments.
- Full and accurate knowledge and understanding of key concepts.
- Detailed awareness of scale and temporal change where appropriate.

AO2

- Detailed evaluative conclusion.
- Detailed application of knowledge and understanding throughout.
- Full evidence of links between knowledge and understanding.

Level 3 (11–15 marks)

AO1

- Generally clear and relevant knowledge and understanding of place(s) and environments.
- Generally clear and accurate knowledge and understanding of key concepts.
- Generally clear awareness of scale and temporal change where appropriate.

AO2

- Clear evaluative conclusion.
- Generally clear application of knowledge and understanding.
- Generally clear evidence of links between knowledge and understanding.

Level 2 (6–10 marks)

AO1

- Some relevant knowledge and understanding of place(s) and environments.
- Some knowledge and understanding of key concepts.
- Some awareness of scale and temporal change where appropriate. There may be a few inaccuracies.

AO2

- Some sense of a conclusion.
- Some partially relevant application of knowledge and understanding.
- Some evidence of links between knowledge and understanding.

Level 1 (1–5 marks)

AO1

- Very limited knowledge and understanding of place(s) and environments.
- Very limited knowledge and understanding of key concepts.
- Very limited awareness of scale and temporal change where appropriate. There may be a number of inaccuracies.

AO2

- Very limited and/or unsupported conclusion.
- Very limited application of knowledge and understanding.
- Very limited evidence of links between knowledge and understanding.

Notes for answers

AO1 Demonstrates clear knowledge and understanding:

- United Nations remit.
- Various agencies of the UN — UNESCO, UNICEF, WHO, World Food Programme, World Bank, UNEP, WMO, IPCC, MDG, SDG, and others.
- The work of these agencies.
- Examples.

A02 Applies knowledge and understanding to offer clear analysis and evaluation:

- Specific examples to highlight the success, or not, of the agency.
- World Bank loans to big prestige projects (e.g. Volta Dam) rather than smaller, bottom-up schemes (e.g. sand dams).
- IMF financial aid to less-developed countries, but limits put on donor country government's spending in education and health, creating further problems.
- WHO eradication of smallpox and polio; AIDS/HIV programmes.
- Peacekeeping not so successful in the short term — Balkans, Middle East, etc.
- UNESCO listing and preservation of world heritage sites, for example Abu Simbel, but not in the Three Gorges area.
- IPCC, Rio, Kyoto — work towards sustainability; Agenda 21 carbon emission targets.

Hints and tips

This is an essay based on the evaluation of a statement, which means looking at it from different sides. As there is a lot of information to cover, a clear structure developed in a brief plan is useful. Illustration with clear examples/case study is essential in order to reach a conclusion. Remember: Start with a statement as to the extent of your agreement, then justify why you feel that way. Finally, make a summative statement based on all that you have said in the answer.

Question 12 example responses

Student A

The United Nations and its various agencies were established in 1945 as a replacement for the League of Nations, originally in order to prevent another world war. In addition, a number of other agencies were established so that the main functions of the UN could be carried out. These include peacekeeping and security, the protection of human rights, economic development and humanitarian assistance, and, more recently, addressing environmental issues. These agencies include UNICEF, UNESCO, the World Bank, the World Health Organization and the World Food Programme. In recent years the UN Environmental Programme, the World Meteorological Organization and the IPCC have been formed to address climate change and global warming, among other things. The UN was also instrumental in drawing up the MDG and SDG.

The work of these agencies has been seen throughout the world, not least the work of the UNHCR. There has not been a world war, but the world has been engaged in war somewhere every year since 1945, except for 1957. As has been seen in the Middle East in recent years, warfare creates large numbers of refugees. The UN through its various agencies has helped support people and provide water and medicines to the sick and elderly in Turkey, to the displaced Rohingya people in Bangladesh, and to people in many parts of Africa such as the DRC, Rwanda and Sierra Leone.

The WHO advises national governments on technical issues relating to health and care systems, particularly on the prevention and treatment of disease. Its efforts have led to the disappearance of some diseases that were quite common, such as smallpox and polio, which is almost gone. Work through the MDG, developed by the WSSD after Rio 1992, has helped improve the health of women and children worldwide but especially in those nations that are less developed.

The Earth Summit in Rio, and the development of Agenda 21 followed by the Kyoto and Paris Agreements, has helped raise the profile of environmental issues, and the IPCC has highlighted the need for drastic action to be taken to address the concerns over global warming and the climate change that must follow. Every member of the UN has now signed up to the Paris Accord, with the notable exception of the USA, showing the success of the UN in promoting international cooperation.

Where the UN has been less successful is in the prevention of war, albeit localised but nevertheless resulting in huge casualty figures and the displacement of large numbers of people. Peacekeepers are deployed and in most cases peace returns, in the Balkans for example, but this may take many years to happen.

The World Bank and the IMF, tasked with providing support for less-developed countries and regulating the world financial systems, have received some criticism. The World Bank makes loans to aid development and thereby reduce poverty, but the conditions attached to the loans often do not have the desired effect. Money is lent for 'top-down' projects like using dams to produce HEP, but in poorer countries this does not help reduce poverty levels. For the last 20 years or so the WB has put more money into sustainable schemes, although there are still many large projects going ahead. The IMF, although not part of the UN, has been known to insist on cuts to government spending in less-developed areas as a condition of getting financial aid. Both these agencies have helped to exacerbate inequality and injustice in parts of the world where this already exists to a large degree.

The conclusion that must be reached is that in preventing a world war and promoting cooperation and development, the UN and its agencies have been successful. It is not until the detail of the levels of success is examined that it becomes clear that the success is not 100%, and in many cases is far from it. Getting every nation to agree to a course of action is success itself but the implementation of that plan still relies on the individual governments, which may have an alternative point of view and programme. The MDG were a great success, as was the eradication of smallpox; the agreement to limit carbon emissions was a partial success, and the prevention of wars has not been at all successful. The statement in the question can be agreed on from both sides: some success but work still to do.

e **The student gives a well-balanced assessment, showing good knowledge and understanding of the issue. The response contains good detail along with supporting evidence. There is a clear structure and an evaluative conclusion.** **Level 4**

Student B

The United Nations was set up to prevent a third world war. In order to do this, it needed to have all the countries of the world agree to work together to remove inequalities and to promote growth and economic and political stability.

The UN has done this by sending peacekeeping troops into areas where there is conflict, such as the Balkans in the 1990s. These troops keep the warring groups apart and try to establish a peaceful environment so that normal life can resume. When the time is right the troops are withdrawn so that the country or countries can continue to develop peacefully on their own. So while it has not stopped wars, the UN has played a role in calming things down and getting them back on an even keel.

One of the biggest problems that the UN has had to deal with is that of sustainability. It has met with some success in this but also some setbacks. The Earth Summit in Rio introduced Agenda 21, a blueprint for the new century intended to make people aware of things like recycling. This was followed 10 years later with the Kyoto summit and, more recently, the Paris summit. The agenda had changed in the meantime and the big concern is now global warming and climate change. All countries have signed up to reduce their carbon emissions, except the USA. So the UN has had almost 100% success but without one of the biggest players!

Money lent by the World Bank was meant to aid development in the poorer member states, thus overcoming the question of poverty. Poverty leads to poor education and healthcare so the loans were very important. Often, though, the money went on prestige projects aimed at making the government look good to its neighbours. Dams to generate HEP were very popular but did little to reduce poverty in the country. Money spent on smaller-scale, bottom-up projects is often far more likely to reduce poverty, especially in rural areas. So, this agency has not been very successful.

The WHO has been far more successful, mainly through the work done by the MDG on women's and children's health. The provision of healthcare systems and advice as well as programmes to educate women have seen a reduction in many childhood diseases. Smallpox has been entirely eradicated and polio has almost disappeared. There is much work still to do with AIDS/HIV, but the programmes in place are meeting with some success.

Agencies like the UNHCR and the WFP have had some success in dealing with the increasing numbers of refugees and with those who have been affected by drought, flood or famine. These are ongoing problems with no end in sight, so it is difficult to say to what degree success has been achieved. Often success in the short term is not repeated in the longer term.

In conclusion, the UN and its agencies have had some success but there are also areas where they have had none or have actually made things worse.

e The student shows some knowledge but expresses it at a simple level. The conclusion is weak, lacking any pulling together of the ideas expressed in the response. In general, more depth and detail are required to achieve Level 4. **Low Level 3**

Section B Changing places

Set 1

Question 1 mark scheme

4 marks (AO1 = 4 marks). 1 mark for each relevant factor named, and 1 mark for the development of a named factor.

Notes for answers

- Choose any two factors from: location, topography, physical geography, land use, built environment and infrastructure, demographic characteristics and economic characteristics.
- Any valid development linking character of place to the named factor.

Question 1 example responses

Student A

Physical geography can contribute to the character of a place. For example, Edinburgh is dominated by Castle Rock and the Old Town was built on it where it could be defended.

Infrastructure also plays a part. Towns built up where there is a major junction of railways, e.g. Swindon and Crewe.

e The student covers two points and shows two development points by way of examples. **4 marks**

Student B

Endogenous factors are ones that come from inside the place. These can include rivers that have a bridge over them.

ⓔ **The student only covers one point. The brief sentence at the beginning is a creditworthy summary of endogenous factors.** **2 marks**

Question 2 mark scheme

6 marks (AO3 = 6 marks). Use the levels of response mark scheme as follows.

- **Level 2 (4–6 marks):** Clear analysis of the data is provided, which makes good use of the evidence to support the points. Clear connections are made between different elements of the data.
- **Level 1 (1–3 marks):** Basic analysis of the data is provided, which makes limited use of the evidence to support points. Basic or limited connections are made between different elements of the data.

Notes for answers

There is a wealth of information here as there are essentially four sets of data in the two figures: Torbay, England, Spitalfields and Banglatown, and Tower Hamlets. Accurate information with supporting evidence will be credited, but comparative statements must also be made. For example:

- Between age groups 0 and 19, Torbay is in line with national trends and shows a narrow base: the very young age group is 2–3% of the population.
- In Torbay there are fewer people in their twenties and thirties than the national average, e.g. for both males and females aged 20–24 there is approximately a 1% difference compared with the national average.
- There are more people in Torbay in the 60+ age groups for both males and females, showing that it has an ageing population with a higher percentage of older people than the national average.
- In both sets of data for the London wards, a large percentage of the population is economically active — nearing 4%. In Torbay this percentage is much lower, e.g. the 19–39 age group averages under 3%.
- There is a much smaller top section in the London data — under 1% in most of the age groups over 60, so there is a much more youthful population in the London areas than in Torbay.
- The base of the pyramid for the London data, e.g. the 0–10 age group, also shows smaller percentages of around 1%. Torbay averages over 2% for the corresponding age groups.

Hints and tips

Analysing data requires you to break down the findings into different parts and then make logical connections using evidence to show the relationship between the different parts.

With four sets of data in this question, there should be plenty of scope for you to make comparative comments.

Question 2 example responses

Student A

The population data for Torbay match those for England in some age groups and not in others. There are many more older people (60+) in Torbay than in England as a whole. For example, for males and females aged 60–64, there is approximately 4% in Torbay and 3% in England, showing that Torbay has an ageing population.

The numbers in the working age groups, e.g. 30–34, are smaller in Torbay (approximately 3%) than in England (approximately 3.8%), showing that Torbay has fewer people of working age. The figures for the lowest age group and the very young, e.g. 0–14, are roughly in line with those of England.

The population data for the two areas in London show a high level in the working population, e.g. 20–30. In Spitalfields and Banglatown, this figure is higher than for England by about 0.5%.

There are fewer old people in the London areas as the population data have a very narrow top.

e The student provides detailed comparative statements with supporting evidence, set out in a clear structure. **Level 2**

Student B

The population pyramid for Torbay has a wide top, showing an ageing population, e.g. the 60+ age groups.

Overall, Torbay has a higher percentage of both males and females than in England in all the age groups from 50+, and a lower percentage of both males and females in the age groups 39 and below.

The data for the London areas show the opposite: there is a narrow top, showing few older people, and a bulging middle, showing a high working population.

The base of both pyramids for the very young, e.g. 0–10, is similar and in line with England, which as a developed country shows a narrow base.

e The student provides comparative statements and covers all the locations, but only in a general way for the London data. The evidence given is in age groups rather than as percentages of the population. However, there is clarity of expression. **Low Level 2**

Question 3 mark scheme

6 marks (AO3 = 6 marks). Use the levels of response mark scheme as follows.

- **Level 2 (4–6 marks):** Clear description of the type of evidence is provided, with clear advantages and disadvantages of the method. Clear links are made between the type of evidence and its outcomes.
- **Level 1 (1–3 marks):** Basic description of the type of evidence is provided with basic strengths and weaknesses. Basic connection(s) are made between aspects of the evidence and the required outcomes.

Notes for answers

Suitable data sources could include:

- statistics, such as census data
- maps
- geo-located data
- geospatial data, including geographic information systems (GIS) applications
- photographs
- text from varied media
- audio-visual media
- artistic representations
- oral sources, such as interviews, reminiscences, songs.

For example:

- A national census is conducted every 10 years. Its strength is that it provides quantitative data that are objective and allow comparisons to be made over time. The census is used for planning purposes by national agencies. It is data collected on a large scale. The weaknesses could be linked to incomplete data collection, especially in places with vulnerable populations.
- Photographs are visual representations that are time and place specific. Their weaknesses are that they are small-scale and dependent on the choice of image of the person taking the photograph. Often the purpose of the photograph was something other than an accurate representation of a place, e.g. the photographs may have been taken for publicity or marketing purposes.

Hints and tips

This question centres on the relative usefulness of each piece of evidence in helping to form perceptions of a place. Remember that you can go further afield than the UK.

Question 3 example responses

Student A

I have chosen census data and photographs.

Census data are regarded as scientific, objective and reliable. These data are very useful in that they provide a large quantity of wide-ranging data that can be analysed and quantified. But, in fact, they are only as reliable as the 'honesty' of the respondents.

Photographs are more open to artistic impression; they are more subjective in terms of both the person taking the photograph and what they want to show, and the person looking at and interpreting the photograph and what they want to see.

Photographs can also be edited and they are very narrow in their range as they show a framed image — right next to this space there may exist something totally contradicting what the image has captured. However, they are useful as they can be analysed and annotated to show a number of things, for example examining the quality of the environment is a good use for photographic evidence.

e **The student gives a balance of the strengths and weaknesses of the two different sources and has related their answer clearly to the usefulness of each resource.** **Level 2**

Student B

Photographs and census data.

Photographs are subjective and census data are objective. Photographs can be edited and 'set up'; census data are more reliable as respondents have to sign to say that their information is accurate. Photographs are considered artistic but a table of statistics is not artistic, it is scientific.

e The student makes a basic comparison with several descriptive but comparative statements. This short answer is limited in depth and detail. Level 1

Question 4 mark scheme

20 marks (AO1 = 10 marks, AO2 = 10 marks). Use the levels of response mark scheme as follows.

Level 4 (16–20 marks)

AO1

- Detailed, relevant and appropriate knowledge and understanding of place(s) and environments.
- Full and accurate knowledge and understanding of key concepts.
- Detailed awareness of scale and temporal change where appropriate.

AO2

- Detailed, evaluative conclusion.
- Detailed application of knowledge and understanding throughout.
- Full evidence of links between knowledge and understanding.

Level 3 (11–15 marks)

AO1

- Generally clear and relevant knowledge and understanding of place(s) and environments.
- Generally clear and accurate knowledge and understanding of key concepts.
- Generally clear awareness of scale and temporal change where appropriate.

AO2

- Clear, evaluative conclusion.
- Generally clear application of knowledge and understanding.
- Generally clear evidence of links between knowledge and understanding.

Level 2 (6–10 marks)

AO1

- Some relevant knowledge and understanding of place(s) and environments.
- Some knowledge and understanding of key concepts.
- Some awareness of scale and temporal change where appropriate. There may be a few inaccuracies.

AO2

- Some sense of a conclusion.
- Some partially relevant application of knowledge and understanding.
- Some evidence of links between knowledge and understanding.

Level 1 (1–5 marks)

AO1

- Very limited knowledge and understanding of place(s) and environments.
- Very limited knowledge and understanding of key concepts.
- Very limited awareness of scale and temporal change where appropriate. There may be a number of inaccuracies.

AO2

- Very limited and/or unsupported conclusion.
- Very limited application of knowledge and understanding.
- Very limited evidence of links between knowledge and understanding.

Notes for answers

AO1 Knowledge and understanding of the attempts made by external agencies to create specific place-meanings. For example:

- External agencies include local government, corporate bodies, tourist organisations, national government, PR companies and private investors.
- Strategies include place marketing, rebranding and reimaging.
- More specific strategies include investment in buildings and services, gentrification processes, top-down investment in large flagship projects, and the location of major sporting, cultural or themed events.
- The focus may be on architecture, retail, art, sport, music or food-related projects or events.

AO2 An evaluation of the impact of creating specific place meanings may focus on the following:

- Environmental improvements — the physical image, new buildings, less derelict land, clean environment.
- Social improvements — lower crime levels, better services, more investment in services and infrastructure due to a growing economy.
- Economic gains — increased employment, increased tax revenue for local government.

An evaluation of its success may be felt differently by different groups of people:

- Locals may feel displaced or be resentful of new social groups moving into an area, for example as the result of gentrification. Increased interest in the area may lead to increased house prices and second-home ownership.
- Visitors may have positive feelings and reimaging may replace out-of-date perceptions of a place.
- People with no physical connection to the place but viewing it as a 'far away' place will be influenced by media rebranding and reimaging.
- The economic impact of a new representation may be the easiest to quantify and discuss.

Hints and tips

The quality of your answer will be determined by the way in which your points are supported by reference to the places you have studied. Two contrasting examples will yield a good amount of discussion and evaluative comments — there is no point choosing two examples that are the same.

'To what extent' is a command term for analysis and evaluation, so evidence is required that reflects the impact of a new representation.

Question 4 example responses

Student A

Rebranding is about marketing a place to give it a new image or new meaning in the view of both locals and outsiders. Reimaging contributes to rebranding and involves the creation of a new image to dispel out-of-date or negative images of a place. It is basically a public relations exercise and may have the aim of attracting investment. Both are strategies used by a range of agencies to influence the meaning or perception of a place.

Many people have a view of Birmingham as an industrial city with remnants of an old industrial past and characteristics of urban decay. Local government in Birmingham has played a vital role in promoting a more positive image in order to attract inward investment to the city.

The strategies involved include new flagship developments such as the construction of a new National Exhibition Centre (NEC). As a result of the city's central location within the country, the NEC can attract major events and exhibitions that will, through the multiplier effect, benefit the whole city and allow more people to visit and leave with a positive image.

The Birmingham Heartlands Development Corporation is a multi-agency body established in 1992 to regenerate the city. Its strategies include new housing projects, facilities and a clean-up of derelict and industrially polluted land.

Birmingham also has a European and International Affairs team to promote the city outside of the UK.

Public spaces and buildings in the city centre have received national acclaim — in the area called Victoria Square, there is an International Convention Centre, a Symphony Hall and the National Indoor Arena. In Centenary Square is the largest public cultural space in Europe. Birmingham City Council sees it as a flagship project for the city's redevelopment. Around Gas Street Basin and Brindley Place, up-market socialising areas provide bars and restaurants for affluent city dwellers.

The example of Birmingham shows that investment by a range of agencies has been used to create a new, more positive and updated place meaning for the city. Its success can be measured in terms of the beneficial knock-on effects to stimulate employment and investment in the city. The upward multiplier effect has created service jobs (48% of the economy). The city's service industry has also benefited from hosting a range of major events such as indoor athletics meetings and the Horse of the Year Show. People travel from all over the country to shop in the refurbished Bullring Centre, which has the flagship stores of high-street leaders such as Selfridges and John Lewis.

On a smaller scale, the rebranding of areas like Notting Hill in London has included gentrification of the area. Conversions of large Georgian and Victorian townhouses have meant that house prices have increased to an average of over £1 million. Place changes such as this can have negative impacts on established residents, who may feel no sense of belonging in the new community.

Birmingham is recognised as a Beta global city. It has the second largest city economy outside of London — there is no doubting the positive knock-on effect of a new representation.

e The student provides detail and place specifics that relate directly to the question and explains how different agencies have sought to alter the place meaning in Birmingham. There is some sense of contrast in scale and impact in the Notting Hill example, but this could have been more detailed. **Level 4**

Student B

Sometimes it is necessary to create a new representation of a place or a new image. It may be that the place has suffered economically and that the negative results of unemployment, poor services and a rundown physical environment need to be reversed in order to create a more positive image. This will attract new investment to raise the economy and make local people feel better and more positive about the place they live in. Sometimes a place may have been badly affected by an environmental event, such as flooding or a tornado, or by a major pollution event or natural disaster, such as an earthquake or tsunami. All of these factors lead to a decline in the image of a place, and external agencies such as local and national governments may become involved to create a new image and kick-start redevelopment.

Belfast as a city became very run down from the 1970s onwards and recently has had a change of image. The city centre has been redeveloped and 'historic quarters' have been established to give a positive sense of its history. Brownfield sites have been redeveloped and the City Council and Tourist Board have worked hard to visually portray a vibrant, cultural city. Seven million tourists have visited Belfast in the last 10 years and this is proof of the positive impacts of its new image. However, it could be said that the period of calm from the unrest and violence of the 1970s onwards has done more to attract people to the city.

Newcastle is another city that has undergone image rebranding to create a new representation of a place. In the 1980s it had many brownfield sites and areas of old industrial development from the port and mining activities. The Tyne and Wear Development Corporation was set up to redevelop brownfield sites, build new housing, clean industrial sites and open flagship building developments such as the Quayside and the Baltic Arts Centre, which was an old flour mill. This has given Newcastle a new, modern and vibrant feel.

In conclusion, external agencies often become involved to create a new place representation.

e The student provides two clear examples of how new place meanings and representations have been created in the context of the question. However, the examples are largely descriptive and the question of to what extent they have been successful is not fully addressed. **Low Level 3**

Set 2

Question 5 mark scheme

4 marks (AO1 = 4 marks). 1 mark for each valid point, with marks available for development.

Notes for answers

- The term 'cultural characteristics' refers to how different cultures develop, regardless of official political/legal boundaries. Examples are when people from a particular place share the same culture (similar values, rituals, religion, food, heroes, symbols, etc.) regardless of the prevailing authority.
- Both in- and out-migration, either international or internal, can play a role. Immigrants tend to congregate in certain places and bring with them their cultural characteristics. They often begin to change the place so that they feel more comfortable in what is to them an alien environment. Out-migration can lead to a cultural void, but this is often filled by the next wave of incomers.

Question 5 example responses

Student A

People move in and out of a town or part of a town. Humans are much more comfortable living close to like-minded people. As soon as there is a group of like-minded people in a small area, that area begins to change and take on its own characteristics. One example, Wilmslow Road in Manchester, has a large number of Indian and Middle Eastern restaurants along the Curry Mile. It contains at least 70 restaurants, take-aways and kebab houses specialising in the cooking of South Asia and the Middle East. There are also a lot of fashion, grocery and music shops along this section of the road.

e The student covers four separate points and gives a clear and accurate explanation. 4 marks

Student B

Cultural characteristics can include religion, shops and restaurants. This has happened in Banglatown where a lot of Bangladeshi immigrants have moved into part of the East End of London and brought shops, restaurants and mosques.

e The student defines cultural characteristics and gives an example. 2 marks

Question 6 mark scheme

6 marks (AO3 = 6 marks). Use the levels of response mark scheme as follows.

- **Level 2 (4–6 marks):** Clear evaluation of the data is provided, which makes good use of the evidence to support the points. Clear connections are made between different elements of the data.
- **Level 1 (1–3 marks):** Basic evaluation of the data is provided, which makes limited use of the evidence to support points. Basic or limited connections are made between different elements of the data.

Notes for answers

The question requires an evaluation of the usefulness of the data. For example:

- Fifteen responses is quite a limited number.
- The time of day of the survey will affect the results, e.g. on a weekday after 9 a.m. and before 3 p.m. there will be no real representation of the working population or of young people of school age. Instead there is likely to be a high proportion of non-working or retired people.
- It is difficult to fully evaluate how useful the data are without knowing what the ten environmental factors were.
- There is a clear numerical difference between the data. The scores are higher for the rural area, indicating a more positive perception of place for the respondents.
- The range for each set of data shows more clustering of the perceptions in the urban population — the urban range is 10 to 42 and the rural range is 15 to 92. The range is much wider for the rural area and the results are more concentrated for the urban area.
- The weather may affect the data scores — people's perceptions could be more negative on a cold, wet day and more positive on a bright, sunny day, so the context of the survey is important.

- It is also difficult to place any confidence in the results for the perception of a 'place' if the survey was just carried out in one small area. All locations or places have good and bad areas, so where the survey was conducted will affect the results. The larger the place, e.g. a large urban area or city, the more unreliable the results. A small village has less scope for differentiation and so, in this way, the scale of the study represents flaws in the data.
- All surveys of this nature are subjective and are open to personal opinion. Therefore direct comparisons and quantified conclusions are difficult to make.

Question 6 example responses

Student A

On the one hand, the data are useful to give a broad overview of the differences in perception of place in an urban and a rural setting. The range of data for each data set is quite different. For the urban setting, the range is 32 (10 to 42) and for the rural setting the range is 78 (15 to 93). This is a useful summary in itself.

However, the data are not useful in that they are from a very small sample — only 15 people. It was carried out only once and we do not know at what time of day. Timing will have a big impact on the people responding, e.g. in the middle of the day there will be a lot of retired people. Also, the exact place will have an impact — urban areas are vast and there will be huge differences in the quality of the environment between different parts of the city.

e The student analyses the strengths and weaknesses of the data within the range of a 6-mark question. Clear knowledge and understanding are shown. **Level 2**

Student B

The data are straightforward and easy to understand. The figures are simple and there is one small data set so it is difficult to see what statistical analysis could be carried out without more data. We do not know what the ten different environmental characteristics were, so the data are limited.

e The student provides a short and limited answer. Basic knowledge and understanding are shown, but evaluation is limited. **Level 1**

Question 7 mark scheme

6 marks (AO1 = 2 marks, AO2 = 4 marks). Use the levels of response mark scheme as follows.

Level 2 (4–6 marks)

- **AO1** — Demonstrates clear knowledge and understanding of concepts, processes, interactions and change.
- **AO2** — Applies knowledge and understanding to the novel situation, offering clear analysis and evaluation drawn appropriately from the context provided. Connections and relationships between different aspects of study are evident, with clear relevance.

Level 1 (1–3 marks)

- **AO1** — Demonstrates basic knowledge and understanding of concepts, processes, interactions and change.

- **AO2** — Applies limited knowledge and understanding to the novel situation, offering basic analysis and evaluation drawn from the context provided. Connections and relationships between different aspects of study are basic, with limited relevance.

Notes for answers

AO1

- There are certain endogenous factors shown in Figure 10:
  - valley-side location
  - steep land on two sides
  - river close by
  - built above the floodplain
  - agricultural area (as well as tourist).
- Figure 11 shows that there is a community spirit with services for members of the community, e.g. a library. Community centres often also encourage adult learning and community events.

AO2

- Endogenous factors seem to dominate in these settlements. Figure 10 tends to highlight the endogenous factors of topography and land use. The building type is fairly uniform, all being stone or white. This could be to do with the national park.
- Similar stone-built properties are shown in Figure 11, but close-up it can be seen that although very rural and quite remote, this settlement is still connected to the wider world.

Hints and tips

The key focus is to get as much out of the photos as possible. It would be useful to show that although the settlements appear remote, they are still connected.

Question 7 example responses

Student A

The photo of Kettlewell shows a small village set on the edge of a steep-sided valley. The fields show that it is an agricultural community, but its location in Wharfedale means it is also a tourist destination. The settlement is nucleated, squeezed between the floodplain of the river and the steep valley sides. Most of the settlement seems old because there are a lot of stone-built houses, though there is a more modern-looking one near the church. There does not appear to be a main road, which means that people living there may feel a bit cut off from the world, especially in winter.

The photo of the Grassington Hub building shows that the village of Grassington offers services to the people who live there, in the form of a library. The fact that it is named the 'hub' shows that it is seen as an important part of the community.

e The student uses both photos and makes reasonable assumptions about the character of the places. Level 2

Student B

Figure 10 shows a small village in the countryside. It is all grouped together round the church and so that could play a big part in the characteristics of the place. It is located in a hilly area with a river in the foreground. There are fields surrounding the village, which make it seem like a farming community.

Figure 11 shows a library. This means that there is at least one service for the local people because people who are there on holiday do not go to the library.

ⓔ **There is some accuracy in the use of Figure 10 but the rest of the answer is simplistic. Level 1**

Question 8 mark scheme

20 marks (AO1 = 10 marks, AO2 = 10 marks). Use the levels of response mark scheme as follows.

Level 4 (16–20 marks)

AO1

- Detailed, relevant and appropriate knowledge and understanding of place(s) and environments.
- Full and accurate knowledge and understanding of key concepts.
- Detailed awareness of scale and temporal change where appropriate.

AO2

- Detailed, evaluative conclusion.
- Detailed application of knowledge and understanding throughout.
- Full evidence of links between knowledge and understanding.

Level 3 (11–15 marks)

AO1

- Generally clear and relevant knowledge and understanding of place(s) and environments.
- Generally clear and accurate knowledge and understanding of key concepts.
- Generally clear awareness of scale and temporal change where appropriate.

AO2

- Clear, evaluative conclusion.
- Generally clear application of knowledge and understanding.
- Generally clear evidence of links between knowledge and understanding.

Level 2 (6–10 marks)

AO1

- Some relevant knowledge and understanding of place(s) and environments.
- Some knowledge and understanding of key concepts.
- Some awareness of scale and temporal change where appropriate. There may be a few inaccuracies.

AO2

- Some sense of a conclusion.
- Some partially relevant application of knowledge and understanding.
- Some evidence of links between knowledge and understanding.

Level 1 (1–5 marks)

AO1

- Very limited knowledge and understanding of place(s) and environments.
- Very limited knowledge and understanding of key concepts.
- Very limited awareness of scale and temporal change where appropriate. There may be a number of inaccuracies.

AO2

- Very limited and/or unsupported conclusion.
- Very limited application of knowledge and understanding.
- Very limited evidence of links between knowledge and understanding.

Notes for answers

AO1 Knowledge and understanding of the past and present connections that shape the place. Connections may originate within a location (endogenous) or from connections, past and present, beyond the locality (exogenous). For example:

- The impact of investment by TNCs.
- The impact of close work or investment from a global institution, e.g. World Bank projects.
- Industrial connections, past and present, e.g. the remnants of an industrial past — mill buildings in the northwest, factory workers' accommodation turned into flats on Tyneside.
- Past and present connections to tourism — seaside towns, e.g. Blackpool.
- The creation of new industrial links, possibly as the result of deindustrialisation and subsequent diversification.
- Place may reflect a strong and continuing connection with the food production industry.
- Migration and urbanisation create external and internal connections that shape a place in terms of the socioeconomic profile and cultural change.

AO2 Application of knowledge and understanding to analyse and evaluate the extent to which located places have been shaped by past and present connections. The exact content will depend on the chosen examples, but shaping places refers to impacts such as:

- the shaping of the built environment — historical and modern forms of housing and industrial buildings
- the shaping of the socioeconomic profile and cultural mix of a place
- the shaping of the economic structure of a place — employment structure, location choice of specific industries, e.g. in the past the Cadbury factory had a significant impact on the physical, social and economic characteristics of Bournville in Birmingham

Hints and tips

A good answer will be determined by either a choice of two examples that allow comparisons to be made or by three detailed and located examples with place-specific facts.

Always start with a statement outlining the extent to which you think shaping has taken place. At the end of the answer you should give a summative statement, making sure that the evidence you have presented matches the extent to which you think shaping has taken place.

Question 8 example responses

Student A

Places are not static; they constantly change over time. Even over a relatively short space of time, such as a year, a major event like a natural disaster or a pollution event can change the physical, social and economic characteristics of a place and change the ways both insiders and outsiders feel about that place, e.g. a flooding event in New Orleans or the destruction from a tornado. This is because places are more than just a location; they are a coming together in time and space of a collection of people and ideas. This essay will show how economic and social connections have shaped the locations of Bournville in Birmingham and Devonport in Plymouth.

Bournville village in Birmingham is a place shaped by the investment and influence of the Cadbury family. Originally a village on the outskirts of Birmingham, it became much more once the Cadbury family decided to expand their chocolate business. Housing and social facilities for the workers were built, including schools, healthcare facilities, homes and public baths. Open spaces were also considered important for the health and welfare of the workers and local community. In 2010 Cadbury was bought by the multinational company Mondelēz. Despite this new, global connection, Bournville — which has been engulfed by the urban growth of Birmingham — still has a 'garden suburb' feel. Large, green spaces remain, planning and building are strictly controlled, and there is a long waiting list for affordable homes. In this way the local connections have changed to global connections in terms of the economic structure of the area, but the physical nature of the area is largely unchanged.

Devonport in Plymouth has changed considerably as a result of external connections. Originally it was a naval dockyard on Plymouth Sound. Naval jobs have declined with military cutbacks and this led to significant social and economic change in the 1980s and 1990s. Levels of deprivation are high in some areas and there are outcomes of increased crime, poor housing and poor social services. Employment continues to decline as the results of the downward multiplier are felt. Recent redevelopment has included physical improvements in housing and the integration of historical landmarks into the redeveloped areas. It is hoped that, as the appearance and image of the place improve, outsiders will be willing to invest and economic development will take place.

These two examples show that places are dynamic and evolve over time. The influences on this development may be from a range of connections that can be social, economic or political in nature. Economic factors tend to have the greatest impact. Sometimes the change to the physical and socioeconomic character of a place can be minimal, but at other times it can be significant and require further intervention.

e The student provides a good introduction, a well-structured and comparative main body with detail from located examples, and a conclusion that reflects the nature of the question. **Level 4**

Student B

Places change as they are not static. When people visit the places they grew up in, they often comment on the nature and extent of change. Places are changed by a range of factors — new people, new investments and new resources. This essay will look at a near and far place in terms of the change each has undergone.

The far place is Medellin, Colombia. For a long time this city was heavily influenced by drugs and violence and it was once known as the most dangerous city in the world. Unemployment, poverty and crime dominated the city. Projects that focus on regeneration and sustainability have given the city a much more positive image. Investment has been made in new education and social services, which help the communities of the city to interact. There is a rapid transport system called Metroplús, which transports people around the city and to employment opportunities. There is also a bike-sharing programme. There are still problems with drugs and violence but not on the previous scale and, although there is less poverty, there is more of a difference between rich and poor. This shows that even with large investments it is sometimes difficult to change a place and its past connections can remain strong.

Devonport in Plymouth has undergone investment to turn its connections with its past industrial dockyard activities into a positive. Old historic monuments have been incorporated into regeneration projects to improve the physical appearance of the town centre. Regeneration has also been aimed at attracting a range of new people to the area with social and shared housing projects. So, in this example, there is a mixture of old and new connections.

Many different external forces can lead to the changes outlined above. Sometimes it is the national government, as in the case of Medellin. In the UK there are various agencies and national government regeneration schemes. TNCs may have an involvement anywhere in the world as they are footloose and can initiate new industrial connections that change a place. Global institutions such as the World Bank can also form a new connection when they finance large-scale projects such as new schools and homes.

e The student shows relevant knowledge and understanding and there is evidence of links between the concepts, e.g. regeneration and reducing poverty. The accounts of the two examples are not detailed enough, however, and there is no clear conclusion. Some idea of the extent of past connections is shown, but again this needs developing. Overall, the information in the essay is limited but would not be described as 'very' limited, so Level 2 is achieved. **Level 2**

Set 3

Question 9 mark scheme

4 marks (AO1 = 4 marks). 1 mark for each valid point, with marks available for development.

Notes for answers

- Maps, road maps, tourist maps — produced by local tourist offices and showing features of interest.
- Statistical sets — population, murders, levels of deprivation, students studying at A-level, etc.
- Advertising copy.
- Film, television, books, plays, songs — with examples.
- Buildings, architecture, iconic buildings, sculptures (e.g. *Angel of the North*).
- Graffiti.

Question 9 example responses

Student A

Any place can be represented in a number of ways, from bare statistics like population figures to references in songs, like Mark Knopfler's 'Silvertown Blues'. Maps can be used to delineate places, show their areal extent and location as well as their proximity to other features. Maps such as the OS Landranger and 1:25,000 series can also highlight specific features.

An important generator of local income is tourism, and advertising copy and tourist material, e.g. 'Catherine Cookson Country', is one way of representing a place or area. The main features of a town are shown and features of lesser interest are left out or minimised.

Places can appear in films or in television series — ITV's *Broadchurch* has done wonders for introducing the area around West Bay to a wider audience in much the same way that *Victoria* has for the Isle of Wight. It is possible to visit *Poldark's* Truro market, which is actually in Wiltshire, and the local town has used this to present and promote itself.

Other art forms such as painting often represent an idealised view of a place, although Lowry's paintings of northern towns give a realistic image of them at the time.

e The student covers several points but there are only 4 marks available for this question. Tighter focus on the question asked would have saved time and still achieved full marks. 4 marks

Student B

Places can be represented in a number of ways. Maps, like the Ordnance Survey maps or road atlases, tell us where places are in relation to other places. Places might be represented as the murder capital of a country according to statistics. And various statistics show that Bognor Regis is one of the UK's worst 'cold spots' when it comes to social mobility.

Places can be represented in film and on television. Sometimes places are nowhere near where they are supposed to be, like some of the locations in *Poldark*. Places can also appear in songs, in paintings or in books.

e The student covers a number of points, but further development or an additional point is needed. 3 marks

Question 10 mark scheme

6 marks (AO1 = 2 marks, AO2 = 4 marks). Use the levels of response mark scheme as follows.

Level 2 (4–6 marks)

- **AO1** — Demonstrates clear knowledge and understanding of concepts, processes, interactions and change.
- **AO2** — Applies knowledge and understanding to the novel situation, offering clear analysis and evaluation drawn appropriately from the context provided. Connections and relationships between different aspects of study are evident, with clear relevance.

Level 1 (1–3 marks)

- **AO1** — Demonstrates basic knowledge and understanding of concepts, processes, interactions and change.
- **AO2** — Applies limited knowledge and understanding to the novel situation, offering basic analysis and evaluation drawn from the context provided. Connections and relationships between different aspects of study are basic, with limited relevance.

Notes for answers

AO1

- The photographs represent two phases of migration, one older and one more recent.
- Local impacts based on fieldwork or personal observation.
- Good opportunity to use work done for independent investigation, if appropriate.

AO2 The impact of flows of people on the cultural characteristics of an area:

- Whitechapel in London, for example — 'smelly' industry, brewing, outside city wall:
  - Seventeenth century: French Huguenot silk makers; French Protestant church; merchants' housing on three and four floors to accommodate looms on top floors.
  - Nineteenth century: Larger numbers of Jews from the Netherlands, Eastern Europe and Russia; church becomes a synagogue; changes to retailing and subdivision of housing; erection of poorer-quality housing to house influx of people.
  - Twentieth century: Bangladeshi migration; synagogue becomes a mosque; further changes to retailing; Brick Lane; Monica Ali.
  - Twenty-first century: Gentrification; changes to retailing.
- Local versus general impacts — Indian restaurants widely used, Eastern European shops serve migrant population but few restaurants, few Asian shops except in specific locations to serve that population, such as Southall in London.
- Examples could come from anywhere in the UK, depending on the student's home area. Liverpool, Birmingham, Leeds and Bradford all display similar patterns. Less appropriate will be small, rural settlements.

Hints and tips

Remember to show that you have used the sources and made a clear appraisal of the usefulness of them.

Question 10 example responses

Student A

The two photographs represent two different stages in the flow of people into an area. The Indian restaurant represents much older flows while the appearance of Eastern European shops is much more recent.

Asian, Indian, Chinese, etc., restaurants can be found in most towns and villages in the UK and they are very much a feature of the British dining experience. However, they have had only a limited impact on the cultural characteristics of an area. The Polish shops supply specific items for the migrant population — items that they are familiar with from home in Warsaw or Vilnius. There are, however, very few Eastern

European restaurants for the host population to try and the host population do not use these foreign shops, preferring their own local shops or supermarkets. Where there is a large Eastern European population, the impact on schools and Catholic churches is significant and host and migrant communities come into closer contact.

In some areas, however, like Whitechapel in London's East End, flows of people have had significant impact. The French Protestants who arrived in the seventeenth century had their own church and built houses to accommodate their silk-making industry. They were replaced from the mid-eighteenth century by flows of Jewish migrants, who came from the economically depressed Netherlands seeking a better way of life. Later in the century, Jews escaping from pogroms in Eastern Europe and Russia added to the numbers and the church became a synagogue. Shops catered for the Jewish population and its religious needs for particular foods. In the twentieth century, an influx of Bangladeshis led to another change — curry houses, sari shops and shops selling Eastern artefacts. And the synagogue is now a mosque! Going to Brick Lane at the weekend is as close to visiting Dakar as many people will get. It is nothing like the average London shopping street, having been completely transformed by the influx of a new population.

What is happening now, as this area is close to the commercial centre of London, is that the houses are becoming gentrified and what were decrepit buildings a few years ago are now fetching huge sums on the property market, changing the cultural characteristics again.

All these changes are confined to a very small area and the impact is watered down as these groups become more assimilated into society and spread out from their initial area of concentration. The two photographs offer a good starting point to discuss the impact that these changes have made, and continue to make, to the cultural make-up of our modern society.

ⓔ **Good knowledge and understanding are demonstrated of the issues embodied in the question. Good examples are given. The student makes well-informed comments, showing an appreciation of the dynamic nature of a place. The response has a good conclusion and evaluative comments at the end. It is, however, very long for a 6-mark question and the time spent on this will impact on how much time the student can spend on other questions. 6 marks**

Student B

Whitechapel in the East End of London is a good example of how different flows of people have changed the cultural characteristics of an area. Originally a poor district, it was settled by French refugees who brought with them the silk-making industry. The houses reflect this, with large workshops on the roofs to get the light. They built their own church in order to worship in the way they wanted.

The French were replaced in the eighteenth and nineteenth centuries by Jews from Europe escaping persecution, and the area's characteristics changed in order to supply the particular needs of the new population. The French church became a synagogue and the silk industry disappeared, to be replaced with the second-hand clothes trade. In more recent times a large Bangladeshi population has made this area of London its home, changing the synagogue into a mosque and replacing the shops selling bagels with curry houses. The area is now being gentrified and another population is moving in, with demands for different shops and more upmarket retailing.

ⓔ **The student shows some knowledge and understanding of one area but ignores the resources and offers only basic analysis and evaluation. The response is largely descriptive. Further evidence and more connections are needed. 3 marks**

Question 11 mark scheme

6 marks (AO3 = 6 marks)

Completing the table (3 marks)

1 mark per correct answer in completing the table. Maximum 3 marks as shown.

Area of London	Observed (O)	Expected (E)	$(O - E)$	$(O - E)^2$	$(O - E)^2 / E$
Northeast	8	5	3	**9**	1.8
Northwest	7	5	2	4	**0.8**
Southwest	2	5	−3	9	1.8
Southeast	3	5	−2	4	0.8
Total	20	20	—	—	**= 5.2**

Interpretation (3 marks)

- The result for the chi-squared calculation of 5.2 falls below the values for the 0.05 and 0.01 significance levels at the 3rd degree of freedom (1 mark).
- Because the value is less than that of the 0.05 significance level, this means that the null hypothesis is accepted and that there is no significant difference in the location of high levels of deprivation in London (1 mark).
- The northeast boroughs have the highest levels of deprivation and the southwest boroughs have the lowest, but this is not significant. (1 mark).

Note: Allow 3 marks for a correct interpretation against an incorrect calculation of chi-squared.

Hints and tips

This question tests your ability to do simple calculations and comment geographically on the result.

Question 11 example responses

Student A

Northeast $(O - E)^2 = 9$; northwest $\frac{(O - E)^2}{E} = 0.8$; total $\frac{(O - E)^2}{E} = 5.2$

The calculation of 5.2 is below the critical value at the 0.05 and 0.01 significance levels, and this means that the null hypothesis is accepted.

It can be stated that there is no significant difference in the location of high levels of deprivation in London.

The student provides three accurate calculations and fully interprets the result with accuracy. **3 + 3 = 6 marks**

Student B

Northeast $(O - E)^2 = 9$; northwest $\frac{(O - E)^2}{E} = 0.8$; total $\frac{(O - E)^2}{E} = 5.2$

From the data, I am able to see that the areas of high deprivation appear to be found in certain parts of London in the NE and NW, but are not so bad in the south and are best in the SW.

e The student provides three accurate calculations but does not complete the task by accepting or rejecting the null hypothesis using Figure 15. Their interpretation is confused and does not relate to the calculation. 3 + 0 marks = 3 marks

Question 12 mark scheme

20 marks (AO1 = 10 marks, AO2 = 10 marks). Use the levels of response mark scheme as follows.

Level 4 (16–20 marks)

AO1

- Detailed, relevant and appropriate knowledge and understanding of place(s) and environments.
- Full and accurate knowledge and understanding of key concepts.
- Detailed awareness of scale and temporal change where appropriate.

AO2

- Detailed evaluative conclusion.
- Detailed application of knowledge and understanding throughout.
- Full evidence of links between knowledge and understanding.

Level 3 (11–15 marks)

AO1

- Generally clear and relevant knowledge and understanding of place(s) and environments.
- Generally clear and accurate knowledge and understanding of key concepts.
- Generally clear awareness of scale and temporal change where appropriate.

AO2

- Clear evaluative conclusion.
- Generally clear application of knowledge and understanding.
- Generally clear evidence of links between knowledge and understanding.

Level 2 (6–10 marks)

AO1

- Some relevant knowledge and understanding of place(s) and environments.
- Some knowledge and understanding of key concepts.
- Some awareness of scale and temporal change where appropriate. There may be a few inaccuracies.

AO2

- Some sense of a conclusion.
- Some partially relevant application of knowledge and understanding.
- Some evidence of links between knowledge and understanding.

Level 1 (1–5 marks)

A01

- Very limited knowledge and understanding of place(s) and environments.
- Very limited knowledge and understanding of key concepts.
- Very limited awareness of scale and temporal change where appropriate. There may be a number of inaccuracies.

A02

- Very limited and/or unsupported conclusion.
- Very limited application of knowledge and understanding.
- Very limited evidence of links between knowledge and understanding.

Notes for answers

One place is sufficient for this question; however, if two places are used then there should be significant contrast. No one place is preferred and this question allows for students to use their own research for the independent investigation to inform their answer.

A01

- Should refer to past and present, even if it is recent past.
- Changing demographic and cultural characteristics in the named place(s).
- Detailed relevant and appropriate knowledge of named place(s).
- Detailed knowledge of change and aspects of continuity in the named place(s).

A02

- An understanding of the scale, possible reasons and rate of temporal, structural and societal change for the named place(s).
- Reference to appropriate statistical data, index of wellbeing, census material and how this contributes to an understanding of change and continuity.
- An understanding of how appropriate map data/photographic data might be used to illustrate the changes/continuity in an area — using a Goad map to show that a shop premises might change usage over time but remain a shop.
- Reference to specific written material — academic and blogs, poetry, paintings, novels that illustrate change and continuity, e.g. Arthur Morrison's *A Child of the Jago,* Monica Ali's *Brick Lane.*
- An understanding that gentrification will have an effect as poorer-quality areas are 'discovered' and gentrified, profoundly changing the lived experience in the area. For example, the rise, decline and rise again of the Spitalfields area of London.
- An understanding that an area adjacent to one that has been gentrified may suffer as a consequence.
- Detailed evaluative conclusion.

Hints and tips

This essay is based on an assessment of a statement. It requires information to be assembled to ultimately agree or disagree with the statement. The question asks for a place that has been studied and without that an answer cannot go beyond Level 1.

An introduction and conclusion are vital in these 20-mark questions and a brief plan at the beginning (in pencil) helps keep an answer focused, making best use of the time available.

Question 12 example responses

Student A

The dynamic character of a place, with changing flows of people against the fluctuations of the economic background, will greatly affect the experience of that place for the people living there.

Wealdstone in northwest London is an area I have studied both through the statistical data available and by interviewing local people. It is part of the London borough of Harrow and although there are areas of Harrow with high levels of affluence, there are also areas such as in Wealdstone with high levels of deprivation. I have also looked at a number of maps and photographs in the library showing change over the last 100 years, both to the built-up area and to the area generally. The local newspaper archive also gives an indication of the changes as well as the continuity of this area.

The town attracted affluent middle-class Londoners from the moment the railway opened in the mid-nineteenth century. The establishment of the Kodak factory and the Whitefriars glassworks drew in a different type of person and the character of the area changed. While the high ground still attracted the affluent who live 'on the hill' near Harrow School, the lower parts like Wealdstone slowly changed.

In the early days, there was a large Irish population and consequently shops, schools and churches that catered specifically for them and added a vibrancy to the area. Catholic churches and Catholic schools still remain in the area although the Irish population has been largely replaced. Shops that sold Irish food, Irish china, and newspapers from every Irish county have largely disappeared as the population has become more integrated, moved back to Ireland or made use of the ease of transport to Ireland to buy particular items. The influx of the Jewish community has brought changes and there is much celebration of the Jewish festivals such as Chanukah, probably with greater enthusiasm than by those living in Israel. But for the long-standing English population this adds another layer of experience to living in the area.

In recent years there has been a large influx of peoples from Asia and the area has the largest population of Sri Lankan Tamils in the UK, as well as a large population of Gujarat Hindus and Jains. Add in Afghans, Syrians, peoples from the Caribbean and Africa, plus more recently EU migrants from Poland and Romania, and it is a far cry from the small, affluent, middle-class housing of the nineteenth century that clustered around the railway station.

These populations bring with them their customs, religions, shops and colour, as well as helping to create a multicultural community. What does remain the same is the average age of the residents, with the early 30s as the median age. This suggests that this is a family-oriented migrant community, with young children offsetting the older members of the host population.

For many of these groups the need to preserve their identity and culture takes the form of shops and community activities; these often focus on the schools and local centres. Over the last hundred years, Wealdstone has seen huge demographic and cultural changes, largely as a result of the changing ethnic composition of the population. This continual change has provided a backdrop to the experience of living in that place. The children who grow up in this environment have a very different experience of place, some would argue a much richer and more rewarding experience, from that of their parents and grandparents. Children in schools learn about the different festivals like Diwali and take part in activities to celebrate them.

Many of the older residents have welcomed the changes, saying that the shops and restaurants are more varied — 'You can eat round the world, a different country every night of the week' — and they have the chance to try different foods and hear about different places. Their experience of place over the 70–80 years they have lived in the area has been one of continuity, but they have also seen the greatest changes, which have enhanced their total experience of place.

The response includes excellent detail and understanding. There is supporting evidence from a case study and a clear conclusion. Level 4: 18 marks

Student B

Wealdstone is in Harrow and is an area that has a lot of different population groups from around the world. First it was the Irish and they had special Irish shops selling what they needed. They also had Catholic schools and churches so they could continue to follow their religion. The Irish integrated well into the community and although there was change it was not great and the area continued to develop.

It was not until large numbers of peoples from India and Sri Lanka arrived that change really happened. The shops started selling different things for the needs of these new migrants and the local people had a different experience of living there as they might not hear English being spoken in the shops they were used to using. This could be a negative experience for people who had lived there all their lives. The increasing numbers of black and minority ethnic groups have radically altered the shopping parades in Wealdstone. These groups have their own churches and religions and celebrate different religious holidays.

However, the local population often joins in with occasions like Diwali, which helps to bring the community together and provides a richness of experience for all.

The experience of place has been affected by the continuous changes that have occurred over the past few years. Wealdstone remains a suburb of London but is a multicultural suburb with peoples from all over the world — Africa, the EU, as well as parts of Asia. All this contributes to the experience of living in the place called Wealdstone.

The student shows some knowledge of the underlying aspects to this question. In general, more depth and detail, along with a better structure, are required for a higher level. The student achieves Level 2 for both AO1 and AO2. Level 2: 8 marks

Section C option: Contemporary urban environments

Set 1

Question 1 mark scheme

6 marks (AO3 = 6 marks). Use the levels of response mark scheme as follows.

- **Level 2 (4–6 marks):** Clear analysis of the data is provided, which makes good use of the evidence to support the points. Clear connections are made between different elements of the data.
- **Level 1 (1–3 marks):** Basic analysis of the data is provided, which makes limited use of the evidence to support points. Basic or limited connections are made between different elements of the data.

Notes for answers

- Most world cities are in the northern hemisphere.
- The highest concentrations are in Europe and on the east coast of the USA.
- Asia has a spread of world cities, mainly in southern and eastern Asia.
- South America and Australasia have some world cities — all in coastal locations.
- Africa as a continent has the fewest world cities and northern Africa has none, despite its close proximity to the markets of Europe.
- There is a general pattern of many world cities being coastal and there are large areas of continental interiors with no representation of world cities, e.g. in the central areas of South America, Africa, Asia and Australasia.

Question 1 example responses

Student A

The map shows a high concentration of world cities in the northern hemisphere, particularly in the northern and central parts of Europe where a number of affluent countries are in close proximity. There is also a large concentration in the USA, particularly on the east coast. The distribution of world cities beyond these concentrations is more widespread. Southern and eastern Asia has several world cities, mainly in coastal locations, illustrating the importance of trade. World cities in the continent of Australasia are mainly on the east coast. In South America there are fewer world cities and they are mainly in the southern part of the continent. Africa is the continent with the fewest world cities and these are mainly in the southern part of the continent. Many continental interiors do not have world cities, e.g. central Asia, Australasia, South America and Africa.

e The student provides detailed evidence of the distribution of world cities using correct geographical terms — northern, southern, continental interiors. Although naming individual countries and cities would give additional detail, this knowledge cannot be assumed as the cities are not named on the map. **Level 2**

Student B

Most world cities are in Europe and the USA. There are very few in Africa and Australasia. In Australia there are five, all of which are on the coast. Asia has a number of world cities, especially in China. South America has ten. There are only a few in Africa and the largest is in South Africa. There are no world cities in the far north and south of the northern and southern hemispheres. Concentrations in Europe dominate the map.

e The student gives some detail of the distribution, but resorts to 'numbers' rather than geographical distribution and spread. Be clear when using 'southern Africa' as a region or 'South Africa' as the country. The structure is not clear as the student returns to concentrations at the end of the response. Ensure you have a logical structure to your answer — the general pattern, high and low concentrations, and then anomalies. Level 1

Question 2 mark scheme

9 marks (AO1 = 4 marks, AO2 = 5 marks). Use the levels of response mark scheme as follows.

Level 3 (7–9 marks)

- **AO1** — The answer shows detailed knowledge and understanding.
- **AO2** — Knowledge and understanding are appropriately applied. Connections and relationships are made between the different aspects studied, which are fully developed and completely relevant. The evaluation is detailed and well supported. The answer provides a well-balanced and coherent argument.

Level 2 (4–6 marks)

- **AO1** — The answer shows clear knowledge and understanding, but there may be some minor inaccuracies.
- **AO2** — Clear knowledge and understanding are appropriately applied. Connections and relationships are made between the different aspects studied, which have some relevance. The evaluation is clear and is supported with appropriate evidence. The answer provides a clear but partial argument.

Level 1 (1–3 marks)

- **AO1** — The answer shows basic knowledge and understanding, with limited relevance, and there are inaccuracies.
- **AO2** — Limited knowledge and understanding are applied. The connections and relationships made between the different aspects studied are basic, with limited relevance. The evaluation is basic. The answer provides a basic argument.

Notes for answers

AO1 Knowledge and understanding of the issues associated with economic inequality. For example:

- An understanding of the concept of economic inequality — the wealth gap between rich and poor in urban areas.
- An appreciation that this can exist over short distances in urban areas and that in poor countries the gap may be substantially wider than in rich countries, despite it being a feature in both.
- Inequality is a relative term — it focuses on the economic status of one group relative to another.

- Issues associated with economic inequality include poverty, unemployment and under-employment, health issues, lack of education, poor housing, lack of mobility and lack of basic services.
- Rich and poor groups become spatially concentrated.

AO2 Evaluation of the issues associated with economic inequality in urban areas. For example:

- Wealthier groups can choose where they live and have the economic ability to live in areas with pleasing environments and good services, such as good quality schools and health facilities. Poorer groups have less choice.
- In areas where there is a low-income population, there is a lower tax revenue (due to a high unemployment percentage) with which to provide local councils with the financial ability to invest in housing, services and infrastructure.
- Disadvantaged areas suffer from a loss of confidence in the area and it is difficult to attract inward investment. This is a result of the downward multiplier effect whereby unemployment leads to a decline in services, out-migration of the economically active, physical decline in the environment, rising crime, lack of confidence and no investment.
- Often, redevelopment and reimaging are required from external sources to break the cycle of economic decline in an area and initiate the upward multiplier process.

Hints and tips

Remember the importance of using subject-specific terms and vocabulary.

Although the question does not directly say 'using examples', examples will always add geographical support to the points you make and are typical of an answer accessing the highest marks.

Your cause-and-effect explanations must show how one thing leads to another.

Question 2 example responses

Student A

Economic inequality is the wealth gap between different groups within urban areas. These can exist over very short distances. Issues vary in rich and poor parts of the world. In the UK, deindustrialisation (the relative decline in importance of manufacturing in urban areas) and decentralising processes (such as suburbanisation and counter-urbanisation) have led to economic differences. When industry relocates, economic decline can result if the industry is not replaced by another economic activity. This is due to the downward multiplier effect. As a result, low-income groups become trapped in rundown areas of a city and are disadvantaged in many aspects of urban living: lack of jobs, poor health and education services, high crime rates due to poverty, barriers to housing and a poor physical environment due to lack of income and investment. Together, these issues combine to form multiple deprivation, which is a measure of inequality.

In low-income countries, like that shown in the photo, the gap between the wealthy and poor can be much greater. The photo shows an urban area where people on good incomes in affluent areas exist right next to shanty towns where rural migrants live. The migrants lack the skills and education required to access jobs and are either unemployed or work in the informal economy. They do not have affordable housing or basic services and amenities and, as a result, a range of social problems exists. The knock-on impacts are poorer health, high unemployment and sometimes social unrest. The poor get stuck in a cycle of poverty from which it is hard to escape. They also become concentrated in poor areas of the city.

e The student shows a good understanding of geographical concepts and explains cause-and-effect sequences. The subject-specific terminology is accurate and is used in the correct context. However, the multiplier effect (upward or downward) is a complex process and should be explained rather than just 'dropped' in to an answer. Level 3

Student B

In poor countries such as that shown in the photograph, rich people live close to shanty towns where there are unemployed people who cannot afford basic housing and services, e.g. clean water and sanitation. This creates health and environmental problems.

Other issues associated with economic inequality include a lack of jobs. Unemployment leads to people having to live in areas where housing is cheaper and the physical environment is poor. Services in health and education are also not as good and this leads to problems with health and being able to get a better job.

In poorer areas, like that shown in the photograph, there is also more crime as people cannot afford everything they need. Because these areas become very run-down, there is often no investment in them and the council cannot afford to do anything and so they get worse and people become trapped.

More affluent people can afford to live and commute to work in areas where there is nice housing and good schools, like in the background of the photo.

e Although the student has a sound general understanding of the issues arising from economic inequality, there could be more use of subject vocabulary, e.g. multiple deprivation, the multiplier effect. There is some evidence of cause-and-effect explanations. Level 2

Question 3 mark scheme

9 marks (AO1 = 4 marks, AO2 = 5 marks). Use the levels of response mark scheme as follows.

Level 3 (7–9 marks)

- **AO1** — The answer shows detailed knowledge and understanding.
- **AO2** — Knowledge and understanding are appropriately applied. Connections and relationships are made between the different aspects studied, which are fully developed and completely relevant. The evaluation is detailed and well supported. The answer provides a well-balanced and coherent argument.

Level 2 (4–6 marks)

- **AO1** — The answer shows clear knowledge and understanding, but there may be some minor inaccuracies.
- **AO2** — Clear knowledge and understanding are appropriately applied. Connections and relationships are made between the different aspects studied, which have some relevance. The evaluation is clear and is supported with appropriate evidence. The answer provides a clear but partial argument.

Level 1 (1–3 marks)

- **AO1** — The answer shows basic knowledge and understanding, with limited relevance, and there are inaccuracies.
- **AO2** — Limited knowledge and understanding are applied. The connections and relationships made between the different aspects studied are basic, with limited relevance. The evaluation is basic. The answer provides a basic argument.

Notes for answers

AO1 Knowledge and understanding of the situation. For example:

- Waste creates a number of issues for urban planners. Inadequate waste disposal leads to air and water pollution.
- Landfill and incineration are the most common forms of waste disposal in high-income countries.
- Space for landfill is running out and incineration is costly.
- Methane from landfills represents 12% of total global methane emissions.
- Recycling rates are increasing but it is clear that much more needs to be done.

AO2 Balancing the advantages and disadvantages. For example:

- Incineration reduces the volume of waste by up to 90%, with the benefit of electricity as a by-product.
- Incineration can sterilise disease agents and reduce the toxicity of waste. The ash can be recycled as aggregate.
- Unregulated burning takes place in low-income countries, in addition to open burning. This leads to air pollution, and dealing with both air pollution and the ash can create significant challenges in poorer countries.
- Landfill in poorer countries involves open dumping; in developed nations it is highly regulated. Again, electricity is a useful by-product and different types of waste are accepted.
- Environmental problems with landfill are numerous, including the fact that methane is produced. Toxic gases can be produced from the bleach and ammonia in landfill, and dust is created.
- Chemicals from landfill can leach into groundwater supplies.
- Landfill is unsightly, takes up space and involves high transport costs.
- Landfill space is running out in many developed nations. Options include taxing landfill to make other options more viable, using the latest incineration technology, which can cut down on carbon dioxide emissions, and using offshore landfill sites.
- Urban mining (reclaiming elements from waste products), trading in waste and e-waste may be also be considered.
- Neither landfill nor incineration has long-term sustainability. The effective long-term sustainable solution incorporates reuse, recycling, reducing and recovery.

Hints and tips

Assessing the effectiveness of both methods (landfill and incineration) requires a consideration of both their strengths and their weaknesses.

Although not directly asked for, your answer should be supported by reference to a located example as specified in the specification content.

Question 3 example responses

Student A

Population growth and economic development are leading to increasing levels of waste. This is particularly so in urban areas where there are high concentrations of people and economic activity. This creates huge challenges for urban managers. Landfill and incineration are the most used waste-disposal techniques, but they have advantages and disadvantages and are used in different ways in rich and poor countries.

Landfill has the advantage of being relatively cheap; it accommodates all types of waste and electricity can be produced as a by-product. However, space for landfill is running out, e.g. as in Amsterdam, and increasingly large transport costs have to be accommodated. The negative side-effects include the production of harmful leachate, which can enter groundwater supplies. Methane gas is also produced and this is a greenhouse gas. These environmental concerns were tackled in Amsterdam where landfill was taxed, encouraging the use of other methods of waste disposal such as incineration.

In low-income countries, landfill is mostly open dumping, which is unsightly, polluting and hazardous to health.

Incineration is costly and air pollution presents a challenge but, again, electricity can be produced as a by-product and incineration reduces the volume of waste by 90%. In Amsterdam, electricity from incineration powers a neighbouring water-treatment plant. Lansink's Ladder is a waste-reduction project that seeks to reduce waste in the first instance by reduction and recycling, incinerating and then dumping only the small amount that is left, in an environmentally friendly way.

On balance, incineration is probably more effective than landfill, but the only real long-term solution is to reduce waste.

e **The student covers the main costs and benefits of incineration and landfill and uses an appropriate example to illustrate their points. There is a clear element of assessing how effective the different methods are as waste-management solutions. Level 3**

Student B

Amsterdam has a comprehensive system for dealing with waste, which includes landfill and incineration. A lack of land and increasing amounts of waste led to a government decision to reduce landfill. The public also objected to landfill sites as they cause air and groundwater pollution and are unsightly. In 1995 landfill was taxed, resulting in a large reduction in the amount of waste going to landfill.

The country has a waste-to-energy scheme where an incineration plant produces electricity and 64% of the waste at the plant is recycled. The energy is used in a water-treatment plant next door.

Landfill and incineration are the main forms of waste disposal in the modern world. In Amsterdam there are different reasons for not using landfill and processing more waste by incineration. Both incineration and landfill deal with waste, but what must be balanced are the advantages and disadvantages. Cost and environmental impacts are clear concerns, but for landfill we have too much waste and too little space. Sometimes rich countries ship waste to other countries to be buried, but this is a limited option.

e **The student addresses both landfill and incineration and recognises the effectiveness of both methods. The answer discusses Amsterdam, which shows good case study knowledge, but there is a danger of being too descriptive if it is not applied well to the question. Level 2**

Question 4 mark scheme

20 marks (AO1 = 10 marks, AO2 = 10 marks). Use the levels of response mark scheme as follows.

Level 4 (16–20 marks)

AO1

- Detailed, relevant and appropriate knowledge and understanding of place(s) and environments.
- Full and accurate knowledge and understanding of key concepts.
- Detailed awareness of scale and temporal change where appropriate.

AO2

- Detailed, evaluative conclusion.
- Detailed application of knowledge and understanding throughout.
- Full evidence of links between knowledge and understanding.

Level 3 (11–15 marks)

AO1

- Generally clear and relevant knowledge and understanding of place(s) and environments.
- Generally clear and accurate knowledge and understanding of key concepts.
- Generally clear awareness of scale and temporal change where appropriate.

AO2

- Clear, evaluative conclusion.
- Generally clear application of knowledge and understanding.
- Generally clear evidence of links between knowledge and understanding.

Level 2 (6–10 marks)

AO1

- Some relevant knowledge and understanding of place(s) and environments.
- Some knowledge and understanding of key concepts.
- Some awareness of scale and temporal change where appropriate. There may be a few inaccuracies.

AO2

- Some sense of a conclusion.
- Some partially relevant application of knowledge and understanding.
- Some evidence of links between knowledge and understanding.

Level 1 (1–5 marks)

AO1

- Very limited knowledge and understanding of place(s) and environments.
- Very limited knowledge and understanding of key concepts.
- Very limited awareness of scale and temporal change where appropriate. There may be a number of inaccuracies.

AO2

- Very limited and/or unsupported conclusion.
- Very limited application of knowledge and understanding.
- Very limited evidence of links between knowledge and understanding.

Notes for answers

AO1 Knowledge and understanding of the concept of urban sustainability — the following should be considered in relation to urban areas:

- An understanding of the varied dimensions of urban sustainability: environmental, social and economic.
- Knowledge of the concept of liveability.
- Knowledge and understanding of the involvement of urban managers — who is responsible for sustainable urban planning?
- An understanding of the two elements of planning: the development of strategies and management, and the implementation of strategies and management.
- Knowledge and understanding of how sustainability is being addressed in specific located examples.
- Knowledge of the features of a sustainable urban area in relation to resource use, e.g. power, food, building materials, waste reduction and recycling, regeneration of land, quality of the built environment, transport issues, use of open spaces, provision of jobs and housing.

AO2 Application of the principles of sustainability to urban areas. For example:

- An analysis of the issues of sustainability: the balance of socioeconomic and environmental considerations, an acceptance of the dynamic nature of urban areas.
- An evaluation of specific examples of sustainable planning.
- A consideration of the approach to sustainability in high- and low-income countries. The growth of most urban areas is in low-income and industrialising countries — how are they approaching sustainable issues?
- The conclusion may acknowledge the fact that the concept of sustainability is idealistic. Urban areas will continue to consume non-renewable resources and pollute the environment, so we have to try to make urban areas more sustainable than they are at present.

Hints and tips

There is an allowance of 40 lines or two pages of exam script for the essay question, so you need to plan your answer so that it is detailed but concise. Make sure you apply the concept of sustainability specifically to urban areas.

Question 4 example responses

Student A

Cities consume three-quarters of the world's resources and as such they present a particular threat to long-term sustainability. The concept of sustainability traditionally means improving quality of life within the Earth's carrying capacity, but for urban areas it has a particular application to make cities safe, inclusive and resilient as well as sustainable. Such a city will provide employment, a high standard of living and a clean, safe and healthy environment, so enhancing its 'liveability'. Examples of green cities include Copenhagen (Denmark) and Curitiba (Brazil). In a model form, green cities recycle and reduce waste outputs, use local produce, conserve energy, use renewable energy and reduce outputs of pollution. They also seek to control widespread urban sprawl. Cities are complex and dynamic and to create a sustainable city is possibly an unachievable aim. This essay will assess some recent attempts to create more sustainable urban areas in an attempt to reduce their ecological footprint.

Urban sustainability is more than just environmental protection; it also includes social development (provision of food, housing and services), economic development (jobs and energy) and urban governance ('green' planning and reducing inequality).

Copenhagen was awarded the European Green Capital of 2014. Its sustainability features include:

- Only 2% of employees work more than 40 hours per week, resulting in more socialising and less stress.
- The city has 249 miles of cycle lanes, which makes transport safe, healthy and environmentally friendly.
- High wages means that there is money from taxes to spend on welfare services and Denmark has a fair and equitable society.

It could be argued that high-income countries have more opportunity to create sustainable areas as they have the financial resources and technology. However, simple sustainable ideas don't need to be costly, e.g. recycling rainwater from gutters or carefully planning housing.

Curitiba in Brazil has a very low-cost, effective method of dealing with waste. In shanty towns where access via narrow streets is difficult, people bring their waste to the collection lorries and receive bus tokens or food in return. The food comes from surpluses from farming and the bus tickets allow them to access jobs in other parts of the city. A lot of the waste-sorting is done by hand, which employs low-skilled workers.

The challenges faced by countries to create sustainable urban areas are huge. Money and good planning are needed. Cities continue to be huge consumers of energy and producers of waste and pollution. Perhaps the most realistic aim is to make cities more sustainable than they are now. Recent successful attempts include the Curitiba Master Plan, the investment in solar panels in Freiberg in Germany, the Groundwork charity in the UK that works to provide more green spaces, and the BedZED development in Greenwich Millennium Village in southeast London.

e The student develops a clear discussion of how achievable urban sustainability is and the reality of this in both rich and poor countries. There is a clear structure and good use of both examples and the correct geographical terminology. High Level 4

Student B

Sustainability means allowing development but ensuring that there are sufficient resources left for future generations to also develop. When applied to a city, it is useful to take a systems approach. Cities have inputs of food, energy and other basic resources that they consume. These are turned into outputs in the living and production processes — waste and pollution are undesirable outputs. In a sustainable city, the inputs will be reduced so as to reduce the carbon footprint of the city, and waste is recycled in a feedback loop. It is the responsibility of urban managers to ensure that the carbon footprint of urban areas is reduced and they become more sustainable.

In practice, this has been achieved in Curitiba in Brazil. Urban waste is collected, sorted and much of it is recycled. Waste is even collected from shanty towns where a bell is rung, people bring out their waste and in return they receive food or bus tokens. The waste-recycling plants also employ lots of people, which creates jobs for low-skilled workers. Also in Curitiba, pollution from vehicles is reduced as they have a cheap and efficient public transport system. Large bi-articulated buses transport people at low cost all over the city, reducing the number of cars. There are also lots of green spaces in Curitiba. When new buildings are planned, builders are allowed to build an extra level in return for allowing open space at ground level. Flooding is reduced as drainage channels and lakes remove excess water when it rains. All of these things reduce the city's negative outputs that are not sustainable, e.g. waste and pollution.

In China, Dongtan eco-city is another attempt to create a sustainable city. It has renewable energy, low-density housing, and is largely car-free as services are within easy reach of housing by cycle lanes. It has no landfill space, so waste is mostly recycled. Jobs are largely contained within the city and provide residents with a comfortable lifestyle. This illustrates the varied nature of an eco-city; it is not just about recycling waste and using renewable energy, it also has to provide a safe and secure way of life with affordable housing for all.

In conclusion, it can be said that examples of eco-cities such as Dongtan and Curitiba show that it is possible to achieve a more sustainable city, but is it possible to achieve a completely sustainable urban area?

e The student recognises the concept of a sustainable city and uses two good case studies. They provide too much descriptive detail about the projects, however — the essay needs to explain what makes these projects sustainable and how they can reflect the aims of creating a truly sustainable urban area. **High Level 2**

Set 2

Question 5 mark scheme

6 marks (AO3 = 6 marks)

Completing the calculation (3 marks)

1 mark for each correct answer in completing the calculation. Maximum 3 marks as shown.

Data collection point	Distance from CBD (km)	Rank	Temperature (°C)	Rank	*d*	d^2
1	0	11	33.5	1	10	100
2	2	10	33	2	8	64
3	4	9	32	3	6	36
4	6	8	28	8	0	0
5	8	7	28.5	6	1	1
6	10	6	30	4	2	4
7	12	5	29	5	0	0
8	14	4	28	8	**–4**	16
9	16	3	28	8	–5	25
10	18	2	26	10	–8	64
11	20	1	25	11	–10	100
						Total = 410

$R_s = 1 - \left(\frac{2460}{n^3 - n}\right)$ $\qquad R_s = 1 - \left(\frac{2460}{1320}\right)$ $\qquad R_s = \mathbf{-0.86}$

Interpretation (3 marks)

- The R_S result of –0.86 exceeds the 0.05 and 0.01 significance levels (1 mark).
- This means that the null hypothesis can be rejected and that there is a significant relationship between distance from the CBD and temperature (1 mark).
- It is a strong negative relationship — as distance increases, temperature decreases (1 mark).

Note: Allow 3 marks for a correct interpretation against an incorrect calculation of R_S.

Hints and tips

This question tests your ability to do simple calculations and comment geographically on the result.

Question 5 example responses

Student A

8	14	4	28	8	**–4**	16

$R_s = 1 - \left(\frac{2460}{n^3 - n}\right)$ $R_s = 1 - \left(\frac{2460}{1320}\right)$ $R_s = \mathbf{-0.86}$

The calculation is above the critical value at both the 0.05 and 0.01 significance levels. This means that the null hypothesis can be rejected and it can be stated that there is a significant relationship between distance from the centre of the city and temperature. It is a negative correlation: as distance increases, temperature decreases.

e The student provides three accurate calculations and fully interprets the result with accuracy. **3 + 3 = 6 marks**

Student B

8	14	4	28	8	**–4**	16

$R_s = 1 - \left(\frac{2460}{n^3 - n}\right)$ $R_s = 1 - \left(\frac{2460}{1320}\right)$ $R_s = \mathbf{-0.86}$

The null hypothesis can be rejected.

e The student produces three accurate calculations and partially interprets the result, but the interpretation is lacking in detail. **3 + 1 = 4 marks**

Question 6 mark scheme

9 marks (AO1 = 4 marks, AO2 = 5 marks). Use the levels of response mark scheme as follows.

Level 3 (7–9 marks)

- **AO1** — The answer shows detailed knowledge and understanding.
- **AO2** — Knowledge and understanding are appropriately applied. Connections and relationships are made between the different aspects studied, which are fully developed and completely relevant. The evaluation is detailed and well supported. The answer provides a well-balanced and coherent argument.

Level 2 (4–6 marks)

- **AO1** — The answer shows clear knowledge and understanding, but there may be some minor inaccuracies.
- **AO2** — Clear knowledge and understanding are appropriately applied. Connections and relationships are made between the different aspects studied, which have some relevance. The

evaluation is clear and is supported with appropriate evidence. The answer provides a clear but partial argument.

Level 1 (1–3 marks)

- **AO1** — The answer shows basic knowledge and understanding, with limited relevance, and there are inaccuracies.
- **AO2** — Limited knowledge and understanding are applied. The connections and relationships made between the different aspects studied are basic, with limited relevance. The evaluation is basic. The answer provides a basic argument.

Notes for answers

AO1 Knowledge and understanding of the challenges. For example:

- There is a need to drain water in urban areas due to the high density of land use, which includes homes, businesses and services.
- There is more water to deal with in urban areas due to the higher levels of precipitation and more intense precipitation.
- Rainfall can result in flash floods due to concrete surfaces and efficient drainage by sloping roofs, gutters and cambered roads.
- Large volumes of water entering rivers at speed combine to present the challenge of potential flooding.
- There are also challenges of pollution from pesticides, pet waste and traffic (grease, salted roads, oil, toxic chemicals).
- If the water system becomes overloaded, raw sewage can surface.

AO2 An assessment of challenges. For example:

- Cost.
- Possible knock-on effects within the catchment area.
- Economic sustainability of the management solutions.
- Opinions and judgements of different stakeholders and the need to reach a consensus on who pays for the solutions.
- The volume and fast flow of water drained from urban areas can create flooding and erosion problems beyond the urban area but within the catchment area, so the challenge is widespread.
- Management attempts include sustainable urban drainage systems (SUDS), which seek to base management on natural catchment processes such as capture and storage.
- The use of located examples to illustrate the challenges.

Hints and tips

The focus of the question is an assessment of the challenges or difficulties of river catchment management in urban areas, so you should make a judgement on the extent and nature of the challenges.

Question 6 example responses

Student A

Urban areas receive more rainfall due to the warmer temperatures (the urban heat island effect), which leads to low pressure. Water drains quickly from the concrete surfaces and through a series of gutters and drains, so infiltration is reduced. Large volumes of fast-flowing water are produced, which then enter streams and rivers — this creates a challenge to management as it can lead to flooding in other areas and increased erosion in the catchment area.

Flooding is extremely costly and disruptive in urban areas due to the mixed land use that includes high-density housing, retail, services, transport and offices. The management of urban river catchments must protect these activities. The photo shows the Los Angeles River flood-control channel. It was cemented and controlled to maximise the flow of water out of the urban area. Such plans can be costly and may create problems further downstream where this water then enters rivers at speed. The challenges of management go beyond flooding and include pollution from pesticides (in parks), pet waste, oil, grease and salt (on roads), all of which can enter groundwater supplies and must be managed. If floodwater overloads the water system, raw sewage can also end up on the surface. The varied nature of the problems requires a varied approach to management. Options include sustainable urban drainage plans that seek a sustainable management solution by using natural processes such as water capture and storage. In Seoul, a major river project has restored the catchment of the urban river that was polluted from urban land use. Three zones of restoration run from the city centre to a conservation area on the edge of the city.

So, the challenges are many, including cost and sustainability, and the problems are of a varied nature.

e The student addresses several of the challenges — large volumes of water, cost, pollution — and makes reference to examples of management. The answer is also brought back to the question focus in a final concluding sentence. Level 3

Student B

Urban areas create a particular challenge in river catchment management as there is a lot of rainfall, which is drained quickly by concrete surfaces and drains. In urban areas there is less evapotranspiration and infiltration, leading to a flash response and high runoff. There is also the addition of wastewater discharge. All of this can quickly enter rivers and streams and lead to flooding in the catchment area. Homes, businesses and services will be flooded and this can cause a lot of upset and disruption.

In rural areas there is more evapotranspiration and infiltration and lower-density land use, and so management is not such a challenge.

More and more natural disasters are flood-related so the challenge is growing. Also, the increased flow of water can lead to problems downstream by causing increased erosion. Therefore, the challenge extends beyond just the urban area. In Los Angeles and Seoul, attempts have been made to manage river catchments by conservation (Seoul) and by forming a concrete river channel as shown in the Los Angeles photograph.

e The student focuses on the flooding and drainage issue and does not recognise the varied challenges relating to pollution and river flow. The answer goes off-track briefly by talking about rural areas, which scores no credit. Examples are referred to but not in any detail. There is some appropriate use of subject vocabulary. Level 2

Question 7 mark scheme

9 marks (AO1 = 4, AO2 = 5 marks). Use the levels of response mark scheme as follows.

Level 3 (7–9 marks)

- **AO1** — The answer shows detailed knowledge and understanding.
- **AO2** — Knowledge and understanding are appropriately applied. Connections and relationships are made between the different aspects studied, which are fully developed and completely relevant. The evaluation is detailed and well supported. The answer provides a well-balanced and coherent argument.

Level 2 (4–6 marks)

- **AO1** — The answer shows clear knowledge and understanding, but there may be some minor inaccuracies.
- **AO2** — Clear knowledge and understanding are appropriately applied. Connections and relationships are made between the different aspects studied, which have some relevance. The evaluation is clear and is supported with appropriate evidence. The answer provides a clear but partial argument.

Level 1 (1–3 marks)

- **AO1** — The answer shows basic knowledge and understanding, with limited relevance, and there are inaccuracies.
- **AO2** — Limited knowledge and understanding are applied. The connections and relationships made between the different aspects studied are basic, with limited relevance. The evaluation is basic. The answer provides a basic argument.

Notes for answers

AO1 An understanding that the main driver of deprivation in urban areas is poverty. General strategies to manage deprivation include:

- Tackling poverty with a minimum living wage.
- Better education, leading to better skills and more employment opportunities.
- Improving the provision of affordable housing.
- Providing affordable public transport.
- Improving the quality of the environment.
- Introducing strategies to reduce crime.
- Attracting investment in industry and services to create jobs.

AO2 An assessment of the difficulties, including:

- A living minimum wage is a central government decision and it may have the knock-on effects of attracting economic migrants and therefore increasing unemployment, or if companies cannot afford the wage, reducing output and employment and leading to practices such as zero-hours contracts.
- Investment in housing, education and transport all require public investment, often at central government level, and this money may not be available.
- A large tax base is required.
- Affordable housing allows the mobility of workers if they need to relocate to find jobs.

Further difficulties may arise from:

- Problems in identifying deprivation.
- Planning and management issues, e.g. assessing which facilities to provide and on what scale.
- The cost implications.
- The nature of the population in the area, e.g. resistance to change, lack of opportunities for self-help.
- The dynamic nature of urban areas — features keep changing.
- The difficulties in attracting private investment to help with new initiatives.

Hints and tips

'Assess' means that rather than just listing management strategies in a descriptive manner, you should give a judgement on the difficulties presented in tackling urban deprivation.

You should refer to located examples to illustrate your points and to access the highest level of marks.

Question 7 example responses

Student A

Urban deprivation exists when a group of social and economic issues combines to leave a section of the population disadvantaged. Managing urban deprivation can be costly and complex as there are many influential factors. The root cause is often poverty caused by job losses. As a result, services decline, the physical environment deteriorates, and there is a loss of confidence in the area and no inward investment.

Providing affordable housing or even renovating existing housing is difficult as the local authorities lack money due to the low tax base from unemployed residents. The economically active often leave, meaning that services are not supported. It is difficult to attract outside investors to help manage the problems, so the deterioration continues. An injection of money is needed to break the cycle of decay. Some management attempts may worsen the situation. For example, offering a minimum wage and training schemes may attract more unemployed people to the area, which stretches services further.

It is also difficult to forecast which services need investment as urban areas are dynamic and the structure of the local population is constantly changing, so money needs to be spent wisely on the right services. It may also be difficult to identify the areas most in need as sometimes not all inhabitants suffer from deprivation and there are many different levels of deprivation.

All of these issues make the management of urban deprivation difficult.

e The student shows a clear understanding of deprivation and its causes and focuses on specific difficulties in addressing the problems and managing effective, long-term change. A good clear structure is shown in this well-explained answer. Level 3

Student B

Urban deprivation exists when people living in urban areas lag behind in several social and economic aspects of life, such as employment, housing quality, services and the quality of the environment. Tackling these issues can be complicated and costly. Also, many different areas are affected. For example, providing better housing requires space and investment. Local authorities in these areas do not have much money and so they are limited in what they can do. The alternative is to attract investment, but this is difficult when the area is rundown and unattractive.

Urban areas are dynamic and so it is difficult to plan long term. Also, because the problems are interlinked, the solutions must also be carefully thought out — there is no point creating new jobs if people don't have the necessary skills, or building new houses if they are not maintained.

ⓔ **The student shows a fair understanding of deprivation and the problems in managing the outcomes. Certain key statements need explaining, e.g. 'local authorities in these areas do not have much money' — why not?** **Level 2**

Question 8 mark scheme

20 marks (AO1 = 10 marks, AO2 = 10 marks). Use the levels of response mark scheme as follows.

Level 4 (16–20 marks)

AO1

- Detailed, relevant and appropriate knowledge and understanding of place(s) and environments.
- Full and accurate knowledge and understanding of key concepts.
- Detailed awareness of scale and temporal change where appropriate.

AO2

- Detailed, evaluative conclusion.
- Detailed application of knowledge and understanding throughout.
- Full evidence of links between knowledge and understanding.

Level 3 (11–15 marks)

AO1

- Generally clear and relevant knowledge and understanding of place(s) and environments.
- Generally clear and accurate knowledge and understanding of key concepts.
- Generally clear awareness of scale and temporal change where appropriate.

AO2

- Clear, evaluative conclusion.
- Generally clear application of knowledge and understanding.
- Generally clear evidence of links between knowledge and understanding.

Level 2 (6–10 marks)

AO1

- Some relevant knowledge and understanding of place(s) and environments.
- Some knowledge and understanding of key concepts.
- Some awareness of scale and temporal change where appropriate. There may be a few inaccuracies.

AO2

- Some sense of a conclusion.
- Some partially relevant application of knowledge and understanding.
- Some evidence of links between knowledge and understanding.

Level 1 (1–5 marks)

AO1

- Very limited knowledge and understanding of place(s) and environments.
- Very limited knowledge and understanding of key concepts.
- Very limited awareness of scale and temporal change where appropriate. There may be a number of inaccuracies.

AO2

- Very limited and/or unsupported conclusion.
- Very limited application of knowledge and understanding.
- Very limited evidence of links between knowledge and understanding.

Notes for answers

AO1 Knowledge and understanding of the environmental issues that pose the greatest challenge to urban managers. For example:

- The key environmental issues affecting urban areas.
- Who the urban managers are, e.g. planners, developers, local and national government.
- The range of environmental issues, e.g. pollution (air, water), urban drainage (flooding, pollution, flows), waste disposal, dereliction.
- Attempts to manage environmental issues in urban areas, with reference to specific examples.
- The wider challenges to urban managers, e.g. economic inequality, poverty, deprivation, social segregation, cultural diversity.

AO2 Evaluation of the environmental issues posing the greatest challenge to urban managers. For example:

- Many environmental issues require immediate action, e.g. pollution events and flooding.
- Many environmental issues are related to further social and economic issues, e.g. derelict land and the cost of regeneration, poor physical environments are an element of urban deprivation, high waste volumes are linked to a consumer-driven society, growing industry produces pollution.
- There is increasing concern over expanding urban areas in low-income and industrialising countries.
- An evaluation of strategies to tackle urban environmental problems such as waste disposal.
- An evaluation of the different approaches in high- and low-income countries, e.g. a lack of money and regulation in low-income and industrialising countries, more awareness and accountability in high-income countries.
- The conclusion may acknowledge the range and complexity of issues facing urban managers, many of which are interlinked.

Hints and tips

There is an allowance of 40 lines or two pages of exam script for the essay question, so you will need to plan your answer so that you are detailed but concise. You are not expected to cover all environmental issues. Always start with a statement of the extent of your agreement. Finish the account with a summative statement.

Question 8 example responses

Student A

Environmental issues that represent a challenge to urban managers include air and water pollution, waste disposal, flooding and dereliction. This is a wide range of issues, which means that the environment poses particularly important challenges to urban managers. Urban areas are constantly expanding and it is likely that these problems will increase. Also, in newly industrialising countries, where most urban growth occurs, many environmental issues are still poorly regulated. This essay will look at two specific environmental problems — air pollution and waste disposal — to illustrate the scale and complexity of the challenge.

Air quality in urban areas can be particularly poor. Sources include carbon monoxide gas (which causes headaches and nausea), nitrogen oxides (which can accelerate weathering processes and cause respiration difficulties) and sulfur dioxide produced in the burning of coal and oil (which causes acid rain and respiration problems).

Reducing the air pollution is extremely important due to its impact on human health. The Clean Air Acts (e.g. in London in 1956) and Air Quality Management Areas (introduced in the 1990s) are examples of past attempts to manage the pollution in the UK. London also has the Ultra-Low Emission Zone, where exhaust emissions are measured. However, air pollution presents particular problems in low-income and industrialising countries where there are few regulations, e.g. in China.

Air pollution presents a particularly important challenge as its effects can spread between countries — a country may receive the negative side-effects of air pollution that it did not cause, which could lead to political tensions. Air pollution is also connected to economic growth, e.g. as cars and industries increase, so do the environmental issues.

Urban waste also presents a significant environmental challenge. The global volume of waste is growing at 7% per year. In addition, waste can also lead to air and water pollution, which impact on human health, e.g. leachates into groundwater supplies. A side-effect is the production of methane gas from landfill sites. This is a greenhouse gas that has a knock-on impact on climate change, which is one of the most important environmental issues of our time. Landfill accounts for 12% of methane gas emissions and 5% of greenhouses gases are produced from waste. Waste generation and its management varies between cities, with low- and middle-income cities showing the lowest levels of management but the highest projected amounts of waste produced. The scale of this issue and its global impact mean that it is one of the greatest challenges to face urban managers. Sources of waste vary within an urban area, making it a significant challenge — sources include household, industrial, commercial and construction.

The challenge of managing urban waste can be turned into a positive with schemes that produce energy from waste, e.g. in Amsterdam an incineration plant powers a water-recycling plant next door.

Although environmental issues present a large-scale problem, there are other problems facing urban managers, such as the social and economic problems that result from unemployment and poverty in cities and lead to the downward multiplier effect — an increase in crime and poor service provision. Also, there are the environmental issues of rundown areas — physical deterioration and derelict land. The environmental challenges are most significant as our environment is a finite resource that must be looked after.

e The student underpins their answer with a detailed understanding of two key environmental problems. It is not possible to cover all environmental problems in sufficient depth, so a choice of two contrasting ones is a good idea. These are clearly set out in the introduction and other environmental and socioeconomic issues are acknowledged. A top-level answer would address the fact that many environmental and social/economic issues are linked, but this is only touched on here. More examples are needed overall. Low Level 4

Student B

Urban change puts increasing pressure on the environment. As cities grow and urban sprawl extends, green environments are taken over by built environments. Urban areas also cause a range of environmental problems, including traffic congestion, air pollution, water pollution, dereliction and waste disposal. These problems create a significant challenge to urban managers.

Chile is a developing country that is facing a range of environmental issues. For a country that does not have a lot of resources, it is facing many environmental problems and struggling to cope. In the city of Santiago, environmental issues pose one of the greatest challenges. The problems include air pollution from cars and growing industries. Santiago is situated in a dip so the pollutants collect in the depression and are trapped there when there is a temperature inversion. This poses a significant challenge as the city's physical geography cannot be changed. There are also high levels of water pollution in Santiago from industries that just dump waste in the river. The problem is that in such countries waste disposal is unregulated. Sewage, chemicals and heavy metals are regularly dumped into the river, posing a significant health hazard.

There is an urban transport management plan to deal with the pollution but it is costly and, in times of recession or when there are social or health problems to manage, the environmental issues often get sidelined.

Some of the wealthiest cities in the world have significant waste problems because increasing affluence means increasing levels of waste are produced. London produces huge volumes of waste and the landfill potential is running out. Although alternatives exist to incinerate more waste or to transport it to different areas, the only real long-term solution is for people to produce less waste. But this is a significant challenge and it involves changing people's habits and attitudes, which is very difficult.

Derelict land is also an issue. It can be linked to deindustrialisation — when industrial land is abandoned, toxic waste and chemicals can be left behind. It is expensive to redevelop these sites and, in times of recession or in low-income countries, such land is often left derelict. It requires a lot of money to clear and redevelop these plots of land.

Environmental problems create a significant challenge as they are very varied in nature and affect all cities. They may not be the most significant challenge, as many cities focus on economic growth and social and cultural improvements before considering the environment.

e The student recognises a range of environmental issues and supports their answer with case study examples. The structure is generally good and there is an introduction and a conclusion. An obvious development of ideas would be to cover the links between environmental and social and economic problems — this is hinted at but not developed. Middle Level 3

Section C option: Population and the environment

Set 1

Question 1 mark scheme

6 marks (AO3 = 6 marks). Use the levels of response mark scheme as follows.

- **Level 2 (4–6 marks):** Clear analysis of the data is provided, which makes good use of the evidence to support the points. Clear connections are made between different elements of the data.
- **Level 1 (1–3 marks):** Basic analysis of the data is provided, which makes limited use of the evidence to support points. Basic or limited connections are made between different elements of the data.

Notes for answers

- Europe has high levels of population density (categories of 100–999 people/km^2) and is a continent achieving 'high' scores in food security.
- India and Bangladesh have high population density (≥1000 people/km^2) but a 'low' score in achieving food security.
- The continent of Africa, excluding for example Egypt in the north and South Africa in the south, largely has a 'very low' score in achieving food security. However, the pattern of population density is varied, with Egypt having a high population density and South Africa a low population density. Much of Africa has very low to low levels of population density.
- China overall has a high population density and a 'medium' score in achieving food security.
- Australasia and North America both have a 'high' score in achieving food security and a very low or low population density.
- Southeast Asia overall has a 'low' score in achieving food security and generally a high population density on most of these small islands.
- As a continent, South America has a 'Medium' score in achieving food security (Venezuela, Ecuador, Peru, Bolivia, Paraguay and Uruguay have a 'low' score) and, overall, fairly low levels of population density.
- Overall, there is evidence of food security correlating negatively with population density, with the exception of Europe. Beyond this, the pattern is very varied and credit is given for analysing the detail of this.

Hints and tips

Always quote figures from the data to support your statements but do not simply make a list. The skill being tested is analysis.

Individual country knowledge is good, but clear references to continents and continental regions, e.g. the northwest of Africa, can equally achieve good analysis marks.

Question 1 example responses

Student A

The maps show some correlation in the areas with highest population density and the areas with a 'high' score in achieving food security, e.g. Japan, but overall the pattern of the two maps is varied. Europe is the only continent where there is a direct correlation of high population density and a 'high' score in achieving food security. The continent of Africa largely has a 'Very low' score in achieving food security, but as a continent it has a varied pattern of population density. In some countries, e.g. Egypt, population density is high and food security is 'Low'. In South Africa, population density is fairly low and food security is 'medium'. However, other countries in the continent, despite having low population density, have a 'very low' score in achieving food security.

The 'high' score areas in terms of food security outside of Europe include North America, Australasia and Japan. North America and Australasia have fairly low population densities and Japan has a high population density.

Overall, there is no clear correlation between population density and food security but there are some areas where there is a negative correlation — a 'high' score in achieving food security and low population density.

e This answer has a clear structure that addresses the overall pattern, highs and lows, and anomalies. The statements are supported by data quoted from the resource. There is a useful conclusion. **Level 2**

Student B

Europe has a high population density and a 'high' score in achieving food security. India has a very high population density but a 'low' score in food security.

Much of Africa has a 'very low' score in food security and relatively low population density.

South America has a 'medium' score in food security and is not overcrowded. China also has a 'Medium' score in food security but with a high population density.

North America and Australasia have 'high' scores in food security and very low or low population densities.

Some areas of Africa in the north and west have a 'low' food security score.

e This answer has a clear structure and quotes figures, but these do not support all the statements. The answer reads as a descriptive list rather than forming a reasoned account of the differences with even basic analysis. **Level 1**

Question 2 mark scheme

9 marks (AO1 = 4 marks, AO2 = 5 marks). Use the levels of response mark scheme as follows.

Level 3 (7–9 marks)

- **AO1** — The answer shows detailed knowledge and understanding.
- **AO2** — Knowledge and understanding are appropriately applied. Connections and relationships are made between the different aspects studied, which are fully developed and completely relevant. The evaluation is detailed and well supported. The answer provides a well-balanced and coherent argument.

Level 2 (4–6 marks)

- **AO1** — The answer shows clear knowledge and understanding, but there may be some minor inaccuracies.
- **AO2** — Clear knowledge and understanding are appropriately applied. Connections and relationships are made between the different aspects studied, which have some relevance. The evaluation is clear and is supported with appropriate evidence. The answer provides a clear but partial argument.

Level 1 (1–3 marks)

- **AO1** — The answer shows basic knowledge and understanding, with limited relevance, and there are inaccuracies.
- **AO2** — Limited knowledge and understanding are applied. The connections and relationships made between the different aspects studied are basic, with limited relevance. The evaluation is basic. The answer provides a basic argument.

Notes for answers

AO1 An understanding that the most important feature of soil is its fertility, i.e. its structure, texture, pH, organic matter and nutrients. Fertility can be maintained by chemicals and fertilisers, but agriculture has to manage a delicate balance in terms of what it takes from the soil and allowing it to recover.

Knowledge of the different problems that lead to soil deterioration:

- Soil erosion by water — soil erosion is the wearing away of the top layer of soil, which is the most nutrient-rich.
- Soil erosion by wind.
- Waterlogging of soil.
- Salinisation.
- Damage to the soil structure.

AO2 Evaluation of the extent to which poor agricultural practices lead to the deterioration of soils. For example:

- Developing the connections between soil deterioration and agriculture.
- Evaluating the extent of the damage caused by agriculture and that caused by other human or natural processes, e.g. drought, fires.
- Acknowledging that sometimes agricultural practices protect and improve soils in harsh environments.
- Assessing the different problems that lead to soil deterioration. Soil erosion can take place by natural processes such as water (rill erosion, sheet erosion, gully erosion and riverbank erosion) and by wind (saltation, creep, abrasion and suspension). However, all of these processes are accelerated if the soil is exposed in vast areas when natural vegetation is removed and agriculture takes over. Cropping leaves soils bare and susceptible to erosion by wind and water.
- Good agricultural practices can mitigate soil erosion, e.g. field drains and ditches channel and remove water, and the sediment can then replace the soil in the fields it came from. Increasing organic matter can improve the cohesion of the soil, crop residues can trap windblown deposits and improve the stability of the soil, increasing plant cover and planting lines of trees as wind breaks can reduce erosion.

- Poor agricultural practices, such as the widespread removal of vegetation and over-cropping, can expose soils to both water and wind erosion.
- If irrigation water exceeds the combined evapotranspiration and percolation rates, waterlogging will result. This means that pore spaces in the soil are occupied by water and not air, and the lack of oxygen affects plant growth.
- An excess of water entering groundwater supplies causes waterlogging. This can be a natural process connected to climatic events or it can be the result of poor agricultural practices if there is enough seepage from irrigation channels.
- Salinisation is the build-up of salts in the soil to a level toxic to plants. It means that plants cannot get enough water from the soil. The process of salinisation is often associated with poor agricultural practices, mainly excessive water application during irrigation. Both groundwater and rainwater contain salts and, when water evaporates, these salts are left behind.
- Salinisation is a particular problem in hot and semi-arid areas where a lot of irrigation water is used, e.g. in China, California and Pakistan there are high rates of evaporation and not enough rainwater to flush the salts out of the soil.
- Agricultural practices are a common cause of the change in the structure of soil. Ploughing and the removal of crops so that little organic matter is returned both affect soil structure.
- Heavy farm machinery compacts soil and makes it impermeable to rainwater. This can lead to gully erosion. A 'plough pan' can form — an impermeable layer in the soil.

Hints and tips

'Assess' requires you to consider the extent to which soil deterioration is caused by poor agricultural practices or by other natural or human processes. Do not be afraid to give alternative views.

Question 2 example responses

Student A

Soil is one of the main natural resources (along with climate) for agriculture. However, poor agricultural practices can damage soils and cause long-term deterioration.

Some processes that lead to soil deterioration include erosion by wind and water, salinisation and structural change. Some of these processes of soil deterioration have natural and human causes and some are caused solely by poor agricultural practices.

Soil erosion occurs when the soil is exposed to wind and water elements. The soil can be removed by rill, sheet and gully erosion (water) or by saltation and abrasion (wind). Although these are natural processes, the exposure of the soil can be the result of poor agricultural practices, e.g. the widespread removal of surface vegetation for agriculture and deforestation. If this practice is not carefully managed, long-term deterioration of soils by erosion will result. Farming practices can reduce erosion and still allow agriculture to take place if practices are put in place, such as digging drainage channels and leaving some crop residues to bind and protect the soil. Wind breaks from lines of planted trees and hedgerows can also reduce erosion — these would represent good agricultural practices. Soil may be exposed due to natural events such as wildfires.

Salinisation occurs when high levels of evaporation in dry climates bring water to the surface. Salts are left behind that become toxic to plants and affect the pH and water balance in the soil. Poor agricultural practices in the form of over-irrigation often lead to salinisation. This is a process that causes deterioration of soil, which can be directly caused by poor agricultural practices.

The structure of soil is often damaged by ploughing. The heavy machinery leaves compacted land that makes the surface impermeable to water, allowing gully erosion to take place, and the soil also has less pore space and less air. A plough pan can develop. This is another example of how poor farming practices can cause soil deterioration.

In conclusion, sometimes soils deteriorate as the direct result of human activity although there can also be natural causes. Good agricultural practices can mitigate and prevent soil deterioration.

e The student covers range of processes that lead to soil deterioration. There is a strong element of evaluation that looks at the extent to which poor agriculture causes this deterioration. **Level 3**

Student B

Soil is important for agriculture. If the surface vegetation is removed in order to plant crops, wind and water will erode the soil, which leads to deterioration as the soil becomes thinner and nutrients are removed. This occurs in areas where the land is farmed intensively and, as continual cropping takes place, more and more nutrients are lost and the soil cannot recover.

Salinisation is a process that destroys soils. Water evaporates and leaves behind salts that, after a period of time, form a hard crust that can also make the soil impermeable. Often water is added by over-irrigation or poor irrigation methods. In hot climates, high levels of evaporation lead to salinisation. The link between irrigation and salinisation is a good example of how poor farming practices lead to soil deterioration.

Ploughing also causes damage to soils. The heavy wheels going over and over the same ground compact the soil and make it impermeable and lacking in open pore spaces for air and water. Again, this shows how poor agricultural practices cause soil deterioration.

e This answer has a good structure and demonstrates knowledge of the key processes that lead to soil deterioration. The evaluation of the extent to which this is caused by poor agricultural practices is not as well developed as in Student A's answer, so the AO2 marks are lower. **Level 2**

Question 3 mark scheme

9 marks (AO1 = 4 marks, AO2 = 5 marks). Use the levels of response mark scheme as follows.

Level 3 (7–9 marks)

- **AO1** — The answer shows detailed knowledge and understanding.
- **AO2** — Knowledge and understanding are appropriately applied. Connections and relationships are made between the different aspects studied, which are fully developed and completely relevant. The evaluation is detailed and well supported. The answer provides a well-balanced and coherent argument.

Level 2 (4–6 marks)

- **AO1** — The answer shows clear knowledge and understanding, but there may be some minor inaccuracies.

- **AO2** — Clear knowledge and understanding are appropriately applied. Connections and relationships are made between the different aspects studied, which have some relevance. The evaluation is clear and is supported with appropriate evidence. The answer provides a clear but partial argument.

Level 1 (1–3 marks)

- **AO1** — The answer shows basic knowledge and understanding, with limited relevance, and there are inaccuracies.
- **AO2** — Limited knowledge and understanding are applied. The connections and relationships made between the different aspects studied are basic, with limited relevance. The evaluation is basic. The answer provides a basic argument.

Notes for answers

AO1 Knowledge and understanding of the challenges of overpopulation. For example:

- The concept of overpopulation — it occurs when there are too many people in an area, putting pressure on the availability of resources.
- An awareness of the implication that only by having fewer people can living standards be raised.
- Reference to the relationship of overpopulation and optimum population, where population and resources are in balance.
- Reference to the concept of carrying capacity, where only certain levels of the consumption of resources can be tolerated.
- Reference to a graphical representation, for example:

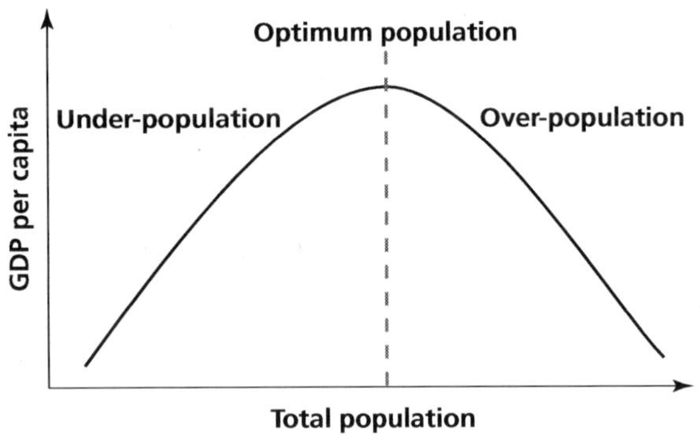

The relationship between population size and standard of living

AO2 Analysis of the extent of challenges. For example:

- Low per capita income and potential poverty due to the large numbers of people and limited supply of resources.
- Unemployment and under-employment.
- Outward migration as people seek a better quality of life elsewhere. Is this a challenge or does it help the government and alleviate the situation? It depends on who is leaving — a 'brain drain', i.e. the loss of highly skilled and economically active sections of society, is a bad thing.
- Insufficient basic resources such as food, minerals, energy and water.

Challenges may be effectively grouped into the following:

- Environmental — increasingly intensive agriculture, deforestation, pollution, increasing waste, habitat loss and desertification as populations try to get more out of the land.
- Economic — high levels of energy use, increased unemployment and price rises as demand for goods exceeds supply; the poverty gap widens.

- Social — housing, health and education services are all compromised.
- Political — conflict; management issues of population and resources, pressure to use protected resources and spaces.

Reference may be made to the following:

- Models such as the Population Resources and Pollution Model and concepts of positive and negative feedback as a management challenge.
- Less obvious examples of under-populated countries, such as UAE and Kuwait, and densely populated countries, such Belgium and Japan. Use these examples to highlight the fact that overpopulation can affect both rich and poor countries, and that the challenges are about the relationship between the numbers of people and the resources available, not just the total population or population density figures on their own.

Hints and tips

The key to this question is that overpopulation is about the *relationship* between the population and the available or sustainable resources.

Question 3 example responses

Student A

Overpopulation exists when there are too many people for the available resources. It is not necessarily dependent on the total population number or on population density, but on the relationship between people and the environment.

Challenges include economic challenges such as how to provide meaningful employment to all people of working age. In a country that is overpopulated, unemployment is a challenge to the government as it can be costly if unemployment benefits must be paid out, and unemployment leads to further social problems such as poverty, poor housing and poor health.

There may also be social challenges such as how to provide services in healthcare, education and welfare for all the people who need them. This will create a huge cost to a government and, unless employment can be maximised, it will not collect enough tax revenue to fund these vital services. The gap between rich and poor may widen according to employment.

There will also be many environmental challenges. Countries such as Mali, where the population is high and the harsh environment is fragile, struggle to support their growing populations in terms of basic food and water needs. As more intense agriculture is practised, environmental damage such as desertification and soil degradation results. A high demand for fuelwood may also put a strain on the environment as the resulting deforestation leaves land exposed to erosion.

Overpopulation can also affect rich countries, e.g. Singapore and Belgium are both classed as overpopulated. However, rich countries may have the capacity to import the things they need, and if they have a high value resource to sell that is in surplus, e.g. oil, they can trade this for shortfalls of resources they do not have.

In conclusion, overpopulation is a complex issue that presents many different challenges depending on the country concerned and its resource base.

e The student considers a range of challenges, grouped under different headings (economic, social, etc.). Good knowledge of the relevant concepts and a clear structure are shown. Level 3

Student B

Overpopulation is when there are too many people for the resources available in a country. A country may also be underpopulated if there are too few people, or it may have an optimum population when people and resources are in balance (this is rare). Challenges include there not being enough jobs for people. This may lead to poverty and civil unrest.

The environment may come under pressure as there is a need to make the best use of what resources are available, e.g. through intensive agriculture, the mining of resources, the overuse of water resources and pressure on energy supplies. In a poor country, fuelwood may be the main source of energy and so this will lead to deforestation. In Brazil the government has been criticised in the past for using the Amazon rainforest for mining, agriculture and farming, but the forest is a resource and the population needs food and jobs.

A Population Resources and Pollution Model was put forward to explain the balance between people and the use of resources. If a country is overpopulated, a further problem may be the waste and pollution produced. This will increase further as there are more and more people — a situation known as positive feedback.

e The student has a clear understanding of overpopulation. A range of material is covered, but there is no detailed analysis of the challenges. Level 2

Question 4 mark scheme

20 marks (AO1 = 10 marks, AO2 = 10 marks). Use the levels of response mark scheme as follows.

Level 4 (16–20 marks)

AO1

- Detailed, relevant and appropriate knowledge and understanding of place(s) and environments.
- Full and accurate knowledge and understanding of key concepts.
- Detailed awareness of scale and temporal change where appropriate.

AO2

- Detailed, evaluative conclusion.
- Detailed application of knowledge and understanding throughout.
- Full evidence of links between knowledge and understanding.

Level 3 (11–15 marks)

AO1

- Generally clear and relevant knowledge and understanding of place(s) and environments.
- Generally clear and accurate knowledge and understanding of key concepts.
- Generally clear awareness of scale and temporal change where appropriate.

AO2

- Clear, evaluative conclusion.
- Generally clear application of knowledge and understanding.
- Generally clear evidence of links between knowledge and understanding.

Level 2 (6–10 marks)

AO1

- Some relevant knowledge and understanding of place(s) and environments.
- Some knowledge and understanding of key concepts.
- Some awareness of scale and temporal change where appropriate. There may be a few inaccuracies.

AO2

- Some sense of a conclusion.
- Some partially relevant application of knowledge and understanding.
- Some evidence of links between knowledge and understanding.

Level 1 (1–5 marks)

AO1

- Very limited knowledge and understanding of place(s) and environments.
- Very limited knowledge and understanding of key concepts.
- Very limited awareness of scale and temporal change where appropriate. There may be a number of inaccuracies.

AO2

- Very limited and/or unsupported conclusion.
- Very limited application of knowledge and understanding.
- Very limited evidence of links between knowledge and understanding.

Notes for answers

The question focuses on the population change section of the specification, specifically the economic implications of migration on both rich and poor countries — whether they are the host country or country of origin.

AO1 Knowledge and understanding of the economic benefits and costs of international migration. For example:

- An understanding of the key ideas of international migration as 'the movement of people across a specified boundary in order to establish a new permanent or semi-permanent residence'.
- International migration affects countries across the development spectrum; mostly the movement is from poor to rich, but movement also exists between rich countries and between poorer countries.
- Migration is the result of a number of factors and will impact both the host country and the country of origin in some way.
- Economic benefits and costs of international migration include: migrants accessing new opportunities; the destination country receiving new skills; the origin country receiving remittance payments; the origin country mostly losing its most productive, skilled workers as they have the finance to travel and the skills required elsewhere.
- Migrants increase the consumption of resources in the host country.
- Economic issues associated with migration include, for example, the effects on jobs, consumption of resources, remittance payments, workers' skills.
- Migration also impacts population structures, which affect dependency ratios in a country.

Points should be supported by reference to case studies and examples.

AO2 Evaluation of the specifically economic impacts of migration on the host and origin countries and how these impact rich and poor countries.

Economic advantages and disadvantages:

- Economic advantages for the country of origin: reduced pressure on consumption of resources; less unemployment; remittances sent home by the migrants; migrants develop new skills that they transfer back home.
- Economic advantages for the host country: specific labour or skills shortages can be overcome; many migrants provide cheap labour; working migrants spend money, which leads to economic growth, and they also pay taxes; an increased workforce can provide economic growth and lead to the knock-on effects of the multiplier; more workers leads to a more favourable dependency ratio.
- Economic costs to the country of origin: often the most skilled and better-educated leave as they have the money and aspirations; there is a dependency on remittances; there is less labour for agricultural and industrial growth; the development of the country may slow down due to a lack of labour.
- Economic costs to the host country: pressure on jobs/unemployment; the economically active who migrate are also in the reproductive age category, which in the long term will increase numbers of the very young, increasing dependency ratios and placing pressure on the need for health and education services.

Analysis of these impacts for rich and poor countries:

- Richer countries will have the capacity to train and find jobs for the migrants, leading to growth, whereas in poorer countries there is not the capacity to invest in education and training.
- Richer countries often benefit from a knock-on effect of more growth and investment due to plentiful, cheap labour.
- Poorer countries lose a large percentage of their small pool of skilled and educated workers.
- The impacts are dependent on scale — small-scale migration with the correct skills set can enhance the development of poorer countries. A large-scale influx of migrants is not economically advantageous for rich or poor countries.
- The overall conclusion may highlight the complexity of the range of potential economic impacts, positive and negative, on host and home country, whether rich or poor.

Case studies and examples are many and varied and may include any of the main current migration routes: Mexicans to the USA; the flood of Syrian and (north) African refugees to Europe; Southeast Asians to North America and the Middle East. Good examples of remittance payments are those to India, China and Mexico.

Hints and tips

A wide range of content is covered by this question, so your essay will benefit from an element of planning to ensure it has a good, clear structure.

Your focus should be on the economic impacts and should not stray into the knock-on social and environmental implications.

There are several variables to consider, e.g. rich/poor/host/home country (or country of origin). Higher-scoring answers will debate how the same factor may have a different impact according to whether the country is rich or poor. The use of examples is not directly asked for but will provide essential support for the points you make.

Question 4 example responses

Student A

International migration is 'the movement of people across a specified boundary in order to establish a permanent or semi-permanent residence'. International migration takes place at a range of scales and for a variety of reasons. Some international migrants are fleeing persecution and are classed as asylum seekers; some are refugees; some are economic migrants who seek employment in another country. This essay will show that all movements will have some degree of economic benefit but the impact will vary across rich and poor countries. In addition to benefits, there are also economic costs for both the host and the home country, whether rich or poor.

Economic benefits for the country of origin include reduced pressure on resource consumption, less unemployment and the remittances (wages) that are often sent home by migrants to their families. These remittances provide spending power in the country of origin, which leads to economic growth.

Economic benefits for the host country include the fact that many migrants provide cheap labour and can transfer useful skills to countries with certain skill deficits. These advantages will be felt in host nations no matter where they are on the development spectrum. However, migrants mostly move from poor to rich countries for economic reasons. Brazil is a country that would be classed as industrialising. It has gained several economic benefits from migration. Migrants to Brazil from Europe have enabled the growth of both the agricultural and the manufacturing sectors. An influx of skilled workers has contributed to the economic development of the country. In addition, Brazilian workers have migrated to Europe and send up to $2.5 billion of remittance payments back home, which allows their families to make home improvements and spend money.

Many rich countries argue that they could not function without foreign workers, e.g. migrant workers make up 85% of the population in the United Arab Emirates, whereas in countries like the USA it is more like 15%. As with other developed countries, migrant workers can often fill the skills gap left in the economy. Often this is low-skilled jobs that the native population may not want to do, e.g. fruit picking and labouring. A knock-on effect in the developed country is that companies can expand and grow due to the supply of cheap labour.

Poor countries on the other hand have more to lose from international migration. They often lose much of their small pool of skilled workers, which delays their own economic development. Haiti and Gambia are examples of countries that have lost more than 50% of their graduates. In Bangladesh, 65% of newly trained doctors seek jobs abroad. On the other hand, remittances boost income and benefit the economy. Africa is a continent that suffers large flows of out-migration due to political unrest and low living standards. In the USA, Canada and Australia, 25–30% of doctors have trained elsewhere.

Alongside the economic gains for both rich and poor countries, there can also be significant economic costs, e.g. shortages of labour in agriculture and industry for low-income countries trying to develop. Low levels of employment lead to low tax revenues and a lack of public spending. High-income countries that accept large numbers of migrant workers may suffer in the long term as migrants are often of reproductive age and therefore in the future the dependency ratio may grow as large numbers of young people will need education and healthcare support.

In conclusion, countries at all stages of the development continuum can experience economic benefits from international migration to some extent, but there are also economic costs. On balance, countries at the lower end of the spectrum will probably experience more economic disadvantages.

e This answer has a good, clear structure. The introduction sets out where the essay is going, the conclusion summarises the viewpoint and each paragraph has a new idea. There is clear knowledge and understanding of a range of issues, supported with facts and an example. **High Level 4**

Student B

Migration is the movement of people to establish permanent or semi-permanent residence in another country. It has costs and benefits for the host country and the country of origin, many of them economic.

Economic advantages for the country of origin include the fact that there will be less unemployment as people who cannot get a job leave to find work elsewhere. If they do not have many skills they will find low-skilled jobs in a more developed country, or it may be that it is the most educated and skilled people who leave and this drains the country of origin of its best workers. An economic advantage is that many workers who leave send money back to their families — known as remittances — and this money helps the economies in poor countries, e.g. Mexico.

The destination country will benefit economically as the migrant workers will often do jobs that the home population do not want to do, e.g. labouring and cleaning. This fills job vacancies and the workers pay taxes and have money to spend. The taxes may just cover the extra demand on services such as healthcare and education, though. There will be economic costs to a host country if migrants do not get jobs and they are in the country long enough to claim benefits. This is a drain on the economy because they are taking benefits and not paying in.

Disadvantages for the country of origin include the fact that, if it is trying to develop, most of its most skilled or young workers will have left. This is true even if the country is trying to develop primary goods. Economic development will be slowed. Sometimes, poor countries attract foreign workers who want a different type of lifestyle or cultural experience. In this case, they will bring new skills to the country and can probably get a high-level job quite quickly and, as it is cheaper in the poor country, they will be attracted by having a much higher personal standard of living than trying to live on a low wage in a city such as London.

So it is not true to say that countries at different ends of the development spectrum benefit. There are advantages and costs to all countries. Also, in a recession, if a country has a lot of migrant workers, it creates bad feeling with locals in need of jobs.

e This answer has a good, basic structure: introduction, main body and conclusion. Although there is a clear understanding of the economic impacts of international migration and some cause and effect is discussed, few geographical examples are given and the essay does not address the full context of the question — it is a straightforward account of advantages and disadvantages to host and origin. **Middle Level 3**

Set 2

Question 5 mark scheme

6 marks (AO3 = 6 marks). Use the levels of response mark scheme as follows.

- **Level 2 (4–6 marks):** Clear analysis of the data is provided, which makes good use of the evidence to support the points. Clear connections are made between different elements of the data.
- **Level 1 (1–3 marks):** Basic analysis of the data is provided, which makes limited use of the evidence to support points. Basic or limited connections are made between different elements of the data.

Notes for answers

Expect supporting detail from quoted figures and some manipulation of the data, e.g. 'twice as many deaths', as percentage figures can be calculated quite quickly. For example:

- There are approximately three times as many deaths by lower respiratory illness in LICs than in HICs (91 – 31 = 60 more deaths per 100,000). In LICs this is the main cause of death and in HICs it is the sixth cause in the table. Of the data shown in the table, respiratory illness accounts for 27% of the deaths in LICs and 8% in HICs.
- Deaths from HIV are high in LICs (accounting for 19% in the data shown), but this illness does not feature in the top six causes of death in HICs.
- There is a high incidence of deaths from strokes in both HICs and LICs, but there are 43 more deaths per 100,000 in HICs, and these account for 23% of the deaths in the data shown for HICs.
- Cancer, Alzheimer's and COPD together account for nearly a third (30%) of the top six causes of deaths in HICs.
- Heart disease accounts for 39% of deaths shown in the table in HICs and 12% in LICs.
- Malaria features as the sixth highest cause of death in LICs but does not feature on the HIC list at all. Diarrhoea and HIV/AIDS also appear on the LIC list but not on the HIC list.
- Cancer, Alzheimer's and COPD appear on the HIC list but not on the LIC list.

Hints and tips

Some simple manipulation of the figures in a set of data can be done quickly and will enhance your answer, e.g. turning raw totals into percentages.

Question 5 example responses

Student A

There is some correlation between the data for HICs and LICs in terms of the diseases in the list, but there are wide differences in the numbers of deaths caused by these diseases.

Lower respiratory illness is first on the list of LICs and sixth on the list for HICs — in LICs there are approximately 60 more deaths per 100,000. Heart disease also features on both lists but, again, there are large differences in the number — 39 per 100,000 deaths (LICs) and 158 per 100,000 deaths (HICs).

There are wide differences in some diseases that appear on one list and not on the other, e.g. in LICs in deaths per 100,000: HIV 65, diarrhoea 53 and malaria 35; and in HICs in deaths per 100,000: cancer 49, Alzheimer's 42 and COPD 31. Together, these three diseases on the HIC list account for almost a third of the deaths in the data.

Overall, the data sets are quite varied.

ⓔ **The student provides detailed comparisons and offers some independent manipulation of the data, e.g. 'a third of the deaths in the data'.** **Level 2**

Student B

The diseases on the two lists are quite different: malaria, diarrhoea and HIV/AIDS on the LIC list, and cancer and Alzheimer's on the HIC list. Most deaths are caused by lower respiratory infections in LICs and these also cause deaths in HICs, but the numbers are 91 and 31 per 100,000, respectively.

Strokes also feature on both lists but, again, the numbers are very different, with 43 per 100,000 more deaths in HICs.

Some diseases in LICs have been eradicated in HICs or can be effectively treated; the diseases in HICs reflect affluent lifestyles.

ⓔ **The student offers some comparative statements with some supporting data. The reasons offered towards the end are not asked for. There needs to be more detailed analysis overall. Level 1**

Question 6 mark scheme

9 marks (AO1 = 4 marks, AO2 = 5 marks). Use the levels of response mark scheme as follows.

Level 3 (7–9 marks)

- **AO1** — The answer shows detailed knowledge and understanding.
- **AO2** — Knowledge and understanding are appropriately applied. Connections and relationships are made between the different aspects studied, which are fully developed and completely relevant. The evaluation is detailed and well supported. The answer provides a well-balanced and coherent argument.

Level 2 (4–6 marks)

- **AO1** — The answer shows clear knowledge and understanding, but there may be some minor inaccuracies.
- **AO2** — Clear knowledge and understanding are appropriately applied. Connections and relationships are made between the different aspects studied, which have some relevance. The evaluation is clear and is supported with appropriate evidence. The answer provides a clear but partial argument.

Level 1 (1–3 marks)

- **AO1** — The answer shows basic knowledge and understanding, with limited relevance, and there are inaccuracies.
- **AO2** — Limited knowledge and understanding are applied. The connections and relationships made between the different aspects studied are basic, with limited relevance. The evaluation is basic. The answer provides a basic argument.

Notes for answers

AO1 Knowledge and understanding of the physical environment and health in the local area. For example:

- The geographical context of the local area and its basic socioeconomic characteristics.
- Specific data on health issues in the local area.
- The physical environmental conditions in the local area.

- The nature of the answer will depend on the local area used:
  - The link between the physical environment and health can have positive or negative impacts.
  - A clean environment with fresh air and opportunities for an outdoor lifestyle and exercise can lead to a healthy lifestyle; an industrial area with air pollution and limited access to open spaces may have negative health implications.

AO2 Application of knowledge to illustrate the specific links between the physical environment and health. For example:

- Climate — high rainfall can lead to damp conditions and more respiratory illnesses. Increased hours of sunshine give health benefits including vitamin D, warmth (can help the elderly with joint conditions), fewer cases of seasonal affective disorder (SAD) and a general feel-good factor that has related health benefits.
- Poor air quality — airborne dust particles or other pollutants from factories/industry may lead to respiratory illnesses, e.g. asthma. High traffic densities also contribute to poor air quality.
- The effects of topography/relief — access to open countryside allows for exercise, which has health benefits related to heart conditions and cholesterol levels. Local topography, such as a dip or valley bottom, however, may also trap pollutants and lead to poor air quality.
- Exercise opportunities are available in urban areas in sports centres so the rural/urban comparison may not be as valid in this respect.
- The relationship between health and the physical environment may not be as strong as the relationship between health and socioeconomic factors.

Hints and tips

The key to this question is to carefully relate physical factors directly to health implications. For example, open, rural spaces allow more opportunities for exercise, which gives health benefits such as reduced cardiovascular illnesses and reduced cholesterol levels.

Question 6 example responses

Student A

The Hart district of Hampshire is one of the most desirable places to live in the UK, as measured in a national poll for four years in a row (2011–14). The district is 42 miles south-southwest of London and it has a large village with a population of 8300. To the east is the River Whitewater valley and to the south open, common land.

Health statistics for the area show that the population has better than the national average scores on a number of health indicators, including a mortality rate lower than the national average and a higher than average life expectancy. Socioeconomic characteristics show low levels of deprivation, high employment and education levels, low obesity and low levels of smoking. The only statistic where levels are higher is road accidents.

Links to the physical environment include a good climate — there is less rainfall than the national average (736 mm per annum against 879 mm) and more sunshine (32.5 hours per week against 29.5 hours). These factors relate to good health as sunshine helps people get out and exercise more and improves vitamin D levels. The air may be less damp so respiratory illnesses such as hay fever and asthma may be reduced.

An attractive physical environment (the New Forest — an AONB) with open countryside and a good climate encourages physical activities that have health benefits in reducing heart disease and improving cholesterol levels; it can also help avoid other health implications of non-communicable disease such as type 2 diabetes.

Although a good physical environment can have positive health benefits as shown, this on its own is not enough and socioeconomic characteristics also have an important role to play. People can be healthy without rural open spaces as urban areas provide parks and sports facilities.

e The student uses a clearly defined local area and relates the physical environment to health in an analytical manner — not just describing facts but linking, for example, climate to specific illnesses and health matters. More figures could have been included to support the facts on the higher life expectancy and lower than national average mortality rates. However, the answer finishes with a good evaluative conclusion. Level 3

Student B

The Hart district of Hampshire is a very desirable place to live. It is classed as a rural area, is 44 miles south of London and has low levels of deprivation.

There are certain environmental factors that mean it is also a healthy place to live. It is over 40 miles from London and so it will not suffer the same atmospheric pollution as London, with its congested roads where the high density of cars gives off carbon monoxide and nitrous gases, both of which cause ill-health such as respiratory illnesses. In the Hart district there are low volumes of traffic and no industry, so air pollution is low and health is good.

The climate is also very good. It has warmer temperatures as it is in the south of the country, which means that there will be fewer colds in the winter. Longer hours of sunshine mean that people will have a feel-good factor that will contribute to good health and also in the winter there will be less seasonal affective disorder (SAD), which can lead to winter depression.

Hart is quite an affluent area, however, so people can afford healthcare, medicines and good food and this also helps people to stay fit and healthy.

e The student uses a located example and provides facts about the physical environment there. The answer is descriptive in places and the facts are not always linked to health benefits — warm winter temperatures and fewer colds is not an accurate link. Overall, more detail is required and more links are needed between health and physical factors. Level 2

Question 7 mark scheme

9 marks (AO1 = 4 marks, AO2 = 5 marks). Use the levels of response mark scheme as follows.

Level 3 (7–9 marks)

- **AO1** — The answer shows detailed knowledge and understanding.
- **AO2** — Knowledge and understanding are appropriately applied. Connections and relationships are made between the different aspects studied, which are fully developed and completely relevant. The evaluation is detailed and well supported. The answer provides a well-balanced and coherent argument.

Level 2 (4–6 marks)

- **AO1** — The answer shows clear knowledge and understanding, but there may be some minor inaccuracies.
- **AO2** — Clear knowledge and understanding are appropriately applied. Connections and relationships are made between the different aspects studied, which have some relevance. The evaluation is clear and is supported with appropriate evidence. The answer provides a clear but partial argument.

Level 1 (1–3 marks)

- **AO1** — The answer shows basic knowledge and understanding, with limited relevance, and there are inaccuracies.
- **AO2** — Limited knowledge and understanding are applied. The connections and relationships made between the different aspects studied are basic, with limited relevance. The evaluation is basic. The answer provides a basic argument.

Notes for answers

AO1 Knowledge and understanding of the features of the demographic transition model. For example:

- The model plots natural population change, including the birth and death rates across time. It is divided into four stages, and countries at differing levels of economic development can be assigned to different stages. A fifth stage has recently been added to the model.
- As a model, it seeks to provide a structure to explain reality.
- The context of the model is important when analysing its present-day weaknesses — it was devised in 1929 and was based on the recorded changes in the birth and death rates of industrialised countries since the industrial revolution. It outlines changes as a country develops from a pre-industrial to an industrial economic system.

AO2 Application of knowledge and understanding to analyse and evaluate the extent to which the model is relevant today. Credit will be given for reference to located example(s). For example:

- The model has to be placed in its historic context in time and place to understand its present-day limitations. As an explanation of the UK as a pioneering industrial nation it has some relevance, but industrialising countries today are developing in a very different context.
- The model does present a starting point for discussion and, in relating it to present-day countries, it does promote a discussion on relevance.
- The main criticism is that the model assumes a 'path' through the stages when, in reality, individual countries will take different transitional routes and may even reverse the path.
- The model considers only economic factors, which is a limitation, but on the other hand this does keep things simple.
- The strengths and weaknesses of the model should not be addressed in a purely descriptive way, e.g. migration is not considered in the model and this has a major impact on many countries today.
- Disease, war and environmental factors are all of relevance today and need to be incorporated into an analysis of any individual country.

Hints and tips

The question asks 'To what extent…', so it is useful to acknowledge some of the strengths of the model and balance these against its limitations in the context of the present day. Reference to an example is needed for a top-level answer.

Question 7 example responses

Student A

It has generally been accepted that the demographic transition model (DTM) is not perfect, as it assumes a path through different set stages and is time- and context-specific. It does, however, provide a useful starting point from which to analyse and compare population change in different countries.

Today, a wide range of physical and human factors play a part in demographic transition. Two examples will be used to illustrate this: Niger and Canada.

Niger is in the Sahel and the physical environment is extremely harsh — recurring drought and desertification are significant physical challenges affecting the population demographics. Most people in rural areas are farmers who have a very difficult existence. Crop failure, insect plagues and internal conflict are further issues. Niger has a high birth rate as a result of religious and cultural beliefs and a low and falling death rate. The death rate is much lower than would be expected in the DTM due to investment and advances in medical care and help from international organisations such as Save the Children.

In Canada there is both a low birth rate and a low death rate; it is a thriving industrial nation with many natural resources. It is in stage 4 of the model and, unlike other industrial nations, has not progressed to stage 5 due to its immigration policy, which means that the population is increasing.

So the DTM is a good starting point for discussion but it has to be accepted that it is out of date in terms of the modern-day factors that it does not refer to.

e The student gives a detailed and well-supported evaluation of the extent to which the demographic transition model can be applied today. Level 3

Student B

The model is useful in many ways as it is easy and straightforward to understand and can be applied as a starting point to many countries. It is possible to assign countries to the different stages in the model, for example Niger (which has a high birth and death rate) and Canada (which has a very low birth and death rate), and countries like Japan where the population is declining.

However, migration is not taken into account and, in the present day, this has a significant impact on a country's population change. Countries all across Europe and especially the richer countries have become a magnet for migrants from poorer countries, as have countries such as Turkey that, due to its proximity to Syria, has taken in thousands of migrants. So in this way the model is not up to date.

War, disease and environmental impacts are other considerations that are not covered in the model but are having a major impact on populations today.

e The student addresses some of the basic strengths and limitations of the model and relates these to present-day factors affecting population trends. There is some reference to different countries, but no depth or detail is given to support the evaluation. The answer is side-tracked by the issue of migration. Level 1

Question 8 mark scheme

20 marks (AO1 = 10 marks, AO2 = 10 marks). Use the levels of response mark scheme as follows.

Level 4 (16–20 marks)

AO1

- Detailed, relevant and appropriate knowledge and understanding of place(s) and environments.
- Full and accurate knowledge and understanding of key concepts.
- Detailed awareness of scale and temporal change where appropriate.

AO2

- Detailed, evaluative conclusion.
- Detailed application of knowledge and understanding throughout.
- Full evidence of links between knowledge and understanding.

Level 3 (11–15 marks)

AO1

- Generally clear and relevant knowledge and understanding of place(s) and environments.
- Generally clear and accurate knowledge and understanding of key concepts.
- Generally clear awareness of scale and temporal change where appropriate.

AO2

- Clear, evaluative conclusion.
- Generally clear application of knowledge and understanding.
- Generally clear evidence of links between knowledge and understanding.

Level 2 (6–10 marks)

AO1

- Some relevant knowledge and understanding of place(s) and environments.
- Some knowledge and understanding of key concepts.
- Some awareness of scale and temporal change where appropriate. There may be a few inaccuracies.

AO2

- Some sense of a conclusion.
- Some partially relevant application of knowledge and understanding.
- Some evidence of links between knowledge and understanding.

Level 1 (1–5 marks)

AO1

- Very limited knowledge and understanding of place(s) and environments.
- Very limited knowledge and understanding of key concepts.
- Very limited awareness of scale and temporal change where appropriate. There may be a number of inaccuracies.

AO2

- Very limited and/or unsupported conclusion.
- Very limited application of knowledge and understanding.
- Very limited evidence of links between knowledge and understanding.

Notes for answers

AO1 Knowledge and understanding of the impact of climate change on health. For example:

- The increased frequency of extreme weather events such as heatwaves, cold spells, floods and droughts.
- Environmental change in terms of the change in temperature and precipitation and the varying pattern of rainfall, which will exacerbate the spread of some diseases and affect water and air pollution levels.
- The agricultural impact on crop yields and the subsequent impact on food security, health and nutrition in developed and developing regions.

AO2 Application of knowledge and understanding to analyse and evaluate the extent of the threat to health from climate change and the geographical spread of the impact. For example:

- Overall, there is a feeling that the impacts of climate change on health will be negative: more severe weather and storms will lead to injuries and fatalities; more air pollution will lead to respiratory illnesses; there will be a potential increase in and widening impact of vector-borne diseases such as malaria and dengue fever; extreme heat will affect the old and young in particular and can cause heat stroke and cardiovascular failure.
- All of the above will have global impacts across countries at all stages of the development spectrum — the differentiating factor will be their ability to prepare and cope.
- There will be some positive impacts — less severe cold in some areas, increased growing seasons in mid to high latitudes.
- In terms of agriculture, areas that are already vulnerable and food insecure will face further crop reductions and water shortages.
- In developed countries, a potential rise in food prices due to supply difficulties could lead to poorer-quality food and increasing obesity.
- There will be more intervention in farming with chemicals and pesticides to improve yields. This will pass through the food chain, harming human health.
- Reducing red meat consumption (i.e. reducing methane gas) will have health benefits from less saturated fat but also health costs resulting from less iron and zinc in people's diets.
- In developing countries there is a link between climate change, declining nutritional standards and migration. Animals are sold when crops fail, which impacts on diet. There may also be an overemphasis on, for example, drought-resistant crops and this will lead to malnutrition and ill health.
- Conclusions may come to the view that the impacts of climate change on health will be far-reaching and largely negative.
- Top-level answers will also cover the extent to which other factors may impact on health, e.g. lifestyle changes, increasing affluence and diet. They will also question whether climate change is 'the biggest health threat'.

Hints and tips

This question covers a lot of potential content and the danger is in just providing a descriptive list of the possible impacts of climate change on health. Instead, your essay should present an argument as to the *extent* to which this may happen. You should refer to the direct and indirect impacts, the extent to which different areas of the world will be affected, and also other possible causes of risks to global health apart from climate change.

Question 8 example responses

Student A

Climate change is predicted to have several key impacts specifically affecting the future health of countries worldwide. The increased frequency of extreme weather events, a change in the patterns of temperature and precipitation, and the impacts on agricultural practices are three important areas where there will potentially be impacts on health. These impacts will be felt in all countries across the development continuum. Climate change will not be the only future health concern; issues such as 'globesity', type 2 diabetes and some cancers all result from unhealthy diet and lifestyle. The question is whether climate change will have the greatest impact.

Climate change will impact health in many possible ways. More severe weather events will cause injuries and fatalities. Increasing events such as heatwaves can have a significant and dangerous impact on the health of the very old and the very young. In Greece and Spain recently, heatwaves led to an increase in fatalities, especially in the older age categories.

Drought will affect the people who already have a vulnerable existence, such as in the Sahel region of Africa. Here, there are already significant problems with meeting food security for all, and more drought will drive down yields and push farmers into even more marginal areas to farm. This will have knock-on health effects of undernutrition, malnutrition and perhaps even starvation.

Methane gas is a greenhouse gas responsible in part for climate change, and meat production contributes to methane levels. Yet reducing meat in diets as a means of controlling greenhouse gases would have health impacts as meat contains iron and zinc, which are needed for good health.

If yields fall and food is in short supply but demand is high, the price will go up. Again, this will impact the health of marginalised groups who already suffer poverty and food insecurity. These groups exist in both rich and poor countries. In rich countries the situation could drive poor people to eat poor-quality food, which can lead to health problems such as obesity.

The pressure on food production may mean more reliance on pesticides and chemicals that seek to raise yields, but these can have dangerous health impacts as they move through the food chain.

There is no doubt that as countries develop there will be a rise in middle-class, more affluent populations. Lifestyles change as a result — in emerging countries more meat is eaten and this has knock-on health impacts of increasing the saturated fat in diets, which leads to cardiovascular illnesses and obesity. There are also major health concerns as some vector-borne diseases are becoming resistant to treatment and for many illnesses antibiotics are not as effective as they used to be. This is a real cause for concern for the future health of populations globally.

In conclusion, there is no doubt that climate change will potentially have a significant and geographically wide-ranging impact on health in the future, but there are also other concerns that should be addressed. Climate change, while significant, is not the only threat to the health of future populations.

e The student shows detailed knowledge and understanding throughout and discusses relationships and connections. There is a clear conclusion that relates to the original question. There could be more exemplification to show geographical place knowledge, but overall this is a well-argued account. Level 4

Student B

Climate change could potentially have major impacts on world health. The main concerns centre on (a) agricultural output and (b) the spread of vector-borne diseases and the effects on people of very hot and very cold spells.

As temperatures warm, more areas will become prone to the impact of disease from vectors, e.g. malaria and dengue fever. Temperature and precipitation are the most important factors to determine the spread of diseases. Studies suggest that climate change could expose an extra 2 billion people outside of the tropics to these diseases.

Zika is a newly emerging disease from infected mosquitoes and it has spread into South and Central America, showing that countries at all levels of development will be affected by the health impacts of climate change.

Very cold and very hot spells will cause health issues, particularly for the very old and very young. With climate change, these weather events will become more frequent and will affect countries worldwide.

The impacts on agriculture will affect food security. An increase in temperatures will push more areas into very dry periods and drought conditions will lead to health problems from undernutrition and malnutrition.

The impacts of climate change on agriculture will be wide-ranging. There will be less rainfall, which means not only a reduction in the quantity of water but also in the quality of water as fresh water is taken for irrigation. There will be an increase in problems such as desertification and salinisation as the land dries out. There may be more weeds and pests as temperatures change, causing crops to fail. Impacts on biodiversity may also wipe out insects that kill pests. All of this will mean less food.

There is consensus now among scientists that climate change is happening and will continue to do so. This essay has shown that change in temperature will have major impacts on vector-borne diseases and agricultural output, which both impact on global health. This has to be a major concern for the twenty-first century.

e This is a well-structured response that addresses the issue with relevant and, on the whole, accurate information (the processes of desertification and salinisation have distinct differences in their causes). There is a paragraph on the agricultural impacts from climate change, but this is not related back to health (e.g. a discussion of undernutrition and malnutrition). Other aspects are also not discussed, such as direct and indirect impacts, impacts on developed and developing regions, and some of the positives of increased temperatures. No conclusion is made either about what other threats there are to global health other than climate change. **Level 3**

